SPACE, TIME AND GRAVITATION

C. Davidson *Frontispiece* *See page 86*

ECLIPSE INSTRUMENTS AT SOBRAL

Arthur S. Eddington

Space, Time and Gravitation

An Outline of the General Relativity Theory

With a Foreword by Dennis Dieks

Edited by Vesselin Petkov

Arthur Stanley Eddington
Born on 28 December 1882 in Kendal, Westmorland, England
Died: on 22 November 1944 in Cambridge, Cambridgeshire, England

Cover: Based on Figure 14 of the book

ISBN: 978-1-927763-30-8 (softcover)
ISBN: 978-1-927763-31-5 (ebook)

Minkowski Institute Press
Montreal, Quebec, Canada
http://minkowskiinstitute.org/mip/

For information on all Minkowski Institute Press publications visit our website at http://minkowskiinstitute.org/mip/books/

Perhaps to move
His laughter at their quaint opinions wide
Hereafter, when they come to model heaven
And calculate the stars: how they will wield
The mighty frame: how build, unbuild, contrive
To save appearances.

Paradise Lost.

Foreword

Eddington is well known for his leading role in the 1919 eclipse expedition, whose results marked the beginning of Einstein's worldwide fame, and for making general relativity known to the English-speaking world. Nevertheless, Eddington's reputation has become somewhat tarnished, due to the unfavorable reception of his later work on "fundamental theory", which has been branded as speculation, and also due to doubts about the objectivity of his work defending Einstein. Regarding the latter, the criticism has first of all been that Eddington, a quaker opposed to nationalism, saw an opportunity to bridge the gap between Germany and its former enemies via the uniting force of international science and thus had a "political" interest in making a biased selection of the eclipse data so as to favor Einstein [3, 11].

In order to judge the case one should know its context and historical background—of which the book *Space, Time and Gravitation* and the other writings republished in the present volume are a central part. Eddington became acquainted with general relativity during the first world war when the astronomer de Sitter—from the neutral Netherlands—sent Eddington, at the time secretary of the Royal Astronomical Society, a number of papers on Einstein's theory. These papers kindled Eddington's interest, and in 1918 he wrote an account of the subject in his *Report on the Relativity Theory of Gravitation* [5]. In his preface Eddington explains why he was attracted to Einstein's theory: it rests on a small number of simple, elegant and universal principles, so that "it claims attention as one of the most beautiful examples of the power of general mathematical reasoning." This statement should be taken seriously: as we shall see, the love of deductive simplicity and elegance is a resounding motif in Eddington's work.

The *Report* was directed at physicists and mathematicians, but in 1920 Eddington followed it up with *Space, Time and Gravitation*, aimed at a wider audience. In this book, reproduced here, Eddington again explains general relativity, but now in a non-technical way, paying extensive attention to conceptual issues and stressing a wider philosophical perspective. Significantly, the book adds to the topics of the 1918 *Report* by including a chapter on the 1919 eclipse expedition.

When one reads *Space, Time and Gravitation* now, one is struck by its high level of sophistication, lucidity and technical competence. Its non-mathematical explanation of general relativity still remains an excellent introduction to the subject—even the connoisseur will encounter illuminating passages. As said, the book's emphasis is on the conceptual and philosophical side, and here one finds several highlights. One example is the Prologue, "What is Geometry?", in which Eddington discusses the status of physical versus mathematical geometry. The account brings to mind the similar one in Reichenbach's highly acclaimed *Philosophy of Space and Time*, which appeared seven years later. However, whereas Reichenbach concludes that physical geometry is conventional and considers the choice between different possible descriptions (with and without "universal forces") as basically arbitrary, Eddington argues that it is the task of physics to describe empirical phenomena without indulging in the addition of superfluous, empirically unsupported, theoretical structure; this fixes a natural geometry. In chapter III, *The World of Four Dimensions*, Eddington offers an account of how four-dimensional spacetime with its Minkowski geometry combines and objectifies all the different "here-and-now" points of view—with a philosophical sophistication one would not expect from a 1920 publication.

There are many other remarkable passages that testify to Eddington's insight, both regarding physics and its philosophy. In one of them Eddington briefly and elegantly analyzes the relation between dynamical and kinematical interpretations of the Lorentz contraction (with a small elaboration in the first Note of the book's mathematical Appendix)—an issue about which even today confusion persists (as shown by debates surrounding Bell's paper [2]). Another of the book's highlights is the discussion of the status of absolute rotation in relativity and of Mach's principle (Chapter X). This chapter could still be used as background reading in a class about relationism and substantivalism with respect to spacetime.

Not unexpectedly, Eddington devotes much space to the question of how light behaves in a gravitational field, as a prelude to his story about the eclipse expedition. The equivalence principle tells us that light *falls* under the influence of gravity, just as ordinary matter. But *exactly how much* will light be bent by material bodies? As Eddington explains in a way that is still enlightening, there are two general relativistic contributions to the effect: one due to a deformation of Euclidean geometry in the presence of masses, and one due to a gravitational effect on time. The magnitude of the latter had already been calculated by Einstein in 1911, a couple of years before the definitive general theory, and turns out to be equal to what Newton's theory of gravitation predicts under the assumption that light possesses mass. The total general relativistic effect is the sum of these two contributions and amounts to twice the value found with Newton's theory. This offers a possibility of putting general relativity to the test of experiment, for example by measuring the deflection of stellar light by the sun. As Eddington writes: "It is this particular test which has turned public attention towards the relativity theory. We shall therefore tell the story of the eclipse expeditions in some detail."

The story is told in Chapter VII. As in the research paper on the subject [4], Eddington here pays ample attention to the reasons that made him and his coworkers select the data as they did (e.g., not all photographs could be considered reliable, in view of a variety of specific circumstances) and to how he came to his conclusion that Einstein's theory accords better with the data than Newton's. This interpretation of the experimental findings by Eddington and his colleagues convinced the scientific community, and the results of the expedition were hailed as a victory of Einstein over Newton.

As already mentioned, it has been objected that the selection of data by Eddington *et al.* was arbitrary and the statistical analysis biased, so that the conclusion in favor of Einstein was not objectively warranted. But as a recent commentator observes, these "criticisms fail to deal with the observers' stated reasons for treating the data as they did, nor do they acknowledge that Eddington *et al.*, as trained professional astronomers, had extensive experience in determining the accuracy and self-consistency of a measurement. Further, the astronomical community, with similar levels of experience and skill, had ample opportunity to check and evaluate their work" [10, p. 88]; see also [1, 7, 8, 9]. Moreover, Eddington's letters to colleagues about his data analysis have been preserved and show how much he was aware of the danger of any *a priori* prejudice and that he took explicit measures to avoid falling into this trap [10, p. 78]. There is no reason to doubt Eddington's sincerity here.

It is an undeniable fact that Eddington employed Einstein's theory as a means to promote internationalism. But that does not imply that he was drawn to the theory because of its possible political implications; it is much more probable that the mathematical beauty and coherence, in addition to the universal scope of the theory, made a decisive impression on him (as we already have seen him declare himself).

In fact, mathematical unification and simplicity played an increasingly important role in Eddington's work. Already in *Space, Time and Gravitation* he expressed the belief that the general theory of relativity could be extended to become a Theory of Everything: a theory that would make it possible to understand the whole universe on the basis of very few simple principles. We should not forget that at this time the only forces that were known in physics were those of electromagnetism and gravitation. Now, in 1918 Hermann Weyl had published his famous gauge theory, which seemed to unite Maxwell's electrodynamics with Einstein's theory of gravitation. This extension of Einstein's theory (in which bodies undergo changes in length when they are transported through regions with electromagnetic potentials) is enthusiastically embraced by Eddington in Chapter XI of *Space, Time and Gravitation.* Even though he notes the objection that this theory entails that the dimensions of, e.g., electrons must depend on their histories, in apparent conflict with experience, the sheer beauty of the theory incites him to theoretical reflections in which we can recognize the outlines of his later "Fundamental Theory".

In Chapter XII Eddington expands these thoughts into a doctrine about "the nature of things". One of the ingredients of this philosophy is remarkably modern: it is the idea that science is only about "structure", namely about the network of *relations* between things, and never about the *essences* of things. Even if there are such essences, these fall outside the scope of science because they are not accessible to us—it is only through our *relations* to physical entities that we acquire knowledge. Similar structuralist ideas are presently a focus of debate in the philosophy of science. The second ingredient in Eddington's philosophy comes more directly from his reflections on relativity theory: it is the idea that matter is not a *cause* of spacetime relations, but rather a *symptom* of them. Usually an equation like $G_{\mu\nu} = T_{\mu\nu}$ is interpreted as saying that the stress-energy tensor of matter $T_{\mu\nu}$ has an effect on the structure of spacetime as expressed by $G_{\mu\nu}$. However, the equation can also be read in the opposite direction, as saying that "material" properties are only an expression of geometrical relations in spacetime.

This line of thought leads Eddington to the idea that the basic things in the world are geometrical: they are "events", in the sense of points in the spacetime continuum. Being elements in a continuum, mathematically speaking these events stand in infinitely many relations to each other. However, some of these relations are more stable and more physically significant than others; in particular, only certain relations generate (via the relation between geometry and matter explained above) stable and lawlike material patterns. Now, Eddington ventures, since we are material ourselves, it is exactly these stable relational structures that we as observers are part of and to which only we respond. In this sense, we ourselves are responsible for the selection of the lawful features of the universe. This idealistic motif explains the famous concluding paragraph of *Space, Time and Gravitation*: "We have found a strange foot-print on the shores of the unknown. We have devised profound theories, one after another, to account for its origin. At last, we have succeeded in reconstructing the creature that made the foot-print. And Lo! it is our own." These thoughts are certainly bold and speculative—although it should be noted that there are similarities to modern "anthropic reasoning" about selection effects. But they also testify to Eddington's relentless desire to *understand* things that usually are taken for granted.

Eddington forcefully propagated his ideas about the philosophical importance of relativity and took part in many debates on the subject. The present volume reproduces two examples of this activity, both brief contributions to discussions published in Nature in 1921. In the first, *The Relativity of Time*, Eddington once again displays his philosophical acumen, among other things by pointing out how deceptive our immediate temporal intuitions are: we think to be aware of a *global now*, but in reality this "global now" is not a matter of perception at all. This brief article is still relevant for contemporary philosophical discussions about the nature of time. In the second piece, *"Space" or "Aether"?*, Eddington continues his thoughts about space, or perhaps better "aether"—geometrical extension being its sole attribute—as the fundamental building block of everything that is physical. Of course, with hindsight this attempt at an all-encompassing Theory of Everything was too premature, and the same can be said of Eddington's later Fundamental Theory. However, this was an honest attempt at understanding the astonishing mathematical order of nature, a theme that is certainly as topical today as it was in Eddington's days.

Summing up, Eddington's writings collected in this volume still provide an excellent non-technical introduction to Einstein's general theory of relativity, with exactly the right amount of detail and concrete examples to give the reader real understanding. Furthermore, and importantly, *Space, Time and Gravitation* and the debates following its publication remain both a milestone in and an indispensable source for the history of relativity theory and its reception—and for the history of twentieth century theoretical physics in general.

Dennis Dieks, Utrecht University

References

[1] Almassi, B. (2009). Trust in expert testimony: Eddington's 1919 eclipse expedition and the British response to general relativity. *Studies in History and Philosophy of Modern Physics*, **40**, 57–67.

[2] Bell, J.S. (2004). How to teach special relativity. Chapter 9 in *Speakable and Unspeakable in Quantum Mechanics*. Cambridge: Cambridge University Press.

[3] Collins, H. and Pinch, T. (1993). *The Golem: What Everyone Should Know About Science*. Cambridge: Cambridge University Press.

[4] Dyson, F.W., Eddington A.S. and Davidson, C. (1920). A determination of the deflection of light by the Sun's gravitational field, from observations made at the total eclipse of May 29, 1919. *Philosophical Transactions of the Royal Society of London A*, **220**, 291–333. Reproduced in [5]

[5] Eddington, A.S. (2014). *Report on the Relativity Theory of Gravitation*; with a foreword by Paul S. Wesson, edited by V. Petkov. Montreal: Minkowski Institute Press. This volume contains a republication of the 1920 (second) edition of Eddington's book plus the eclipse report [4].

[6] Eddington, A.S. (2016). *The Mathematical Theory of Relativity*; with a foreword by Abhay Ashtekar, edited by V. Petkov. Montreal: Minkowski Institute Press. A republication of the second (1924) edition.

[7] Kennefick, D. (2009). Testing relativity from the 1919 eclipse—a question of bias. *Physics Today*, March 2009, 37–42.

[8] Kennefick, D. (2012). Not only because of theory: Dyson, Eddington and the competing myths of the 1919 eclipse expedition. In C. Lehner, J. Renn, and M. Schemmel (eds), *Einstein and the Changing Worldviews of Physics*, 201–232. Heidelberg: Springer.

[9] Petkov, V. (2014). Editorial preface to [5].

[10] Stanley, M. (2003), An expedition to heal the wounds of war: The 1919 eclipse and Eddington as Quaker adventurer. *Isis*, **94**, 57–89.

[11] Waller, J. (2002). *Einstein's Luck: The Truth Behind Some of the Greatest Scientific Discoveries*. Oxford: Oxford University Press.

Editor's Preface

This volume contains new publications of A. S. Eddington's famous book *Space, Time and Gravitation: An Outline of the General Relativity Theory* [1], written for a wider audience, and two short pieces originally published in *Nature* – an article on time (The Relativity of Time [2]) and a Letter to the Editor on space ("Space" or "Aether"? [3]). The short *Nature* publications are included in the volume because they shed additional (and still important today) light on some aspects of the (then) new views of space, time and gravitation.

The volume begins with an excellent Foreword by Dennis Dieks which will be found very informative not only by the general readers but also by physicists and philosophers.

Two things justify the new publication of Eddington's book, which was published in 1920, only four years after the publication of Einstein's general relativity.

1. Eddington is well-known, not only in his day, as a committed and very successful popularizer of science. This is true particularly for Einstein's general relativity because in 1918 Eddington gave its first systematic exposition *Report on the Relativity Theory of Gravitation* [4] and was instantly recognized as one of the most trusted experts in this new and difficult field. Eddington's mastery in writing clearly about challenging and abstract concepts, in full display even in this technical treatise, is the reason why his books are still valuable today. Here is what the renowned astrophysicist Subrahmanyan Chandrasekhar, who received the 1983 Nobel Prize for Physics, wrote in 1983 [5]:

> Eddington's *Report* is written so clearly and yet so concisely that it can be read, even today, as a good introductory text by a beginning student.

Two years later (in 1920) Eddington published *Space, Time and Gravitation: An Outline of the General Relativity Theory* specifically written for non-experts to explain the new and revolutionary ideas about space, time and gravitation. The following year – in 1921 – in a review in the *Bulletin of the American Mathematical Society* Edwin Bidwell Wilson wrote that Eddington's book "is undoubtedly the best general presentation" of general relativity [6]. This view was further strengthened in 1923 when Eddington published his comprehensive treatise on the mathematical and physical foundations of general relativity *The Mathematical Theory of Relativity* [7] because it demonstrated to everyone that the popular book *Space, Time and Gravitation* is written not only by a skilled popularizer of science but also by one of the three (as the legend has it) experts on general relativity.

2. What justifies the new publication not only of Eddington's *Space, Time and Gravitation* but also of his *Nature* pieces are Eddington's views on some fundamental issues that have not yet been resolved. Here are several examples:

2.1. In 2018 we will mark the 110th anniversary of Hermann Minkowski's groundbreaking lecture "Space and Time" [8] in which he presented Eisnstein's special relativity as a theory of flat spacetime (Minkowski called it "die *Welt*" or the [four-dimensional] World), but there is still no consensus among physicists and philosophers on the issue of the reality of spacetime. In the distant 1920 and 1921 Eddington directly confronted this issue:

> The question is often raised whether this four-dimensional space-time is real, or merely a mathematical construction; perhaps it is sufficient to reply that it can at any rate not be less real than the fictitious space and time which it supplants. ["The Relativity of Time," in this volume p. 160]

> However successful the theory of a four-dimensional world may be, it is difficult to ignore a voice inside us which whispers "At the back of your mind, you know that a fourth dimension is all nonsense." I fancy that that voice must often have had a busy time in the past history of physics. What nonsense to say that this solid table on which I am writing is a collection of electrons moving with prodigious speeds in empty spaces, which relatively to electronic dimensions are as wide as the spaces between the planets in the solar system! What nonsense to say that the thin air is trying to crush my body with a load of 14 lbs. to the square inch! What nonsense that the star-cluster, which I see through the telescope obviously there now, is a glimpse into a past age 50,000 years ago! Let us not be beguiled by this voice. It is discredited. [*Space, Time and Gravitation*, in this volume p. 41]

> In a perfectly determinate scheme the past and future may be regarded as lying mapped out—as much available to present exploration as the distant parts of space. Events do not happen; they are just there, and we come across them. [*Space, Time and Gravitation*, in this volume p. 37]

2.2. The issue of the nature of relativistic length contraction (of a rod) has been debated for years (even recently), whereas Eddington clearly argued in favour of Minkowski's explanation of length contraction [8, p. 116] – that the spaces of two observers in relative motion intersect the rod's four-dimensional worldtube at different angles and the resulting cross-sections are of different length; the cross-section of the rod's worldtube for the observer, with respect to whom the rod moves, is shorter. Eddington's position on the nature of length contraction is unambiguously stated:

> The real rod in nature is the four-dimensional object shown in section as $P'PQQ'$. [*Space, Time and Gravitation*, in this volume p. 40]

2.3. The issue of whether relativity got rid of the ether has also been debated for years. In his Letter to the Editor of *Nature* "Space" or "Aether"? Eddington insisted that space is not emptiness (one may argue that, logically, emptiness implies non-existence), but is something that is not three-dimensional and in order to be consistent with general relativity "it is called by Minkowski's term *world*" (in this volume p. 164).

2.4. In his *Nature* article "The Relativity of Time" Eddington makes a bold statement – "gravitation as a separate agency becomes unnecessary" (in this volume p. 161) – which may be interpreted in a sense that gravitation is not a physical interaction since all gravitational phenomena may be fully explained as manifestations of the non-Euclidean geometry of spacetime. Such a stunning interpretation may offer an explanation of the unsuccessful attempts to create a theory of quantum gravity, which does not appear to have been examined so far.

Montreal, 24 August 2017

Vesselin Petkov
Minkowski Institute

REFERENCES

1. A. S. Eddington, *Space, Time and Gravitation: An Outline of the General Relativity Theory* (Cambridge University Press 1920)
2. A. S. Eddington, The Relativity of Time, *Nature* **106**, 802-804 (17 February 1921)
3. A. S. Eddington, "Space" or "Aether"? *Nature* **107**, 201 (April 14, 1921)
4. A. S. Eddington, *Report on the Relativity Theory of Gravitation* (Fleetway Press, London 1918); new publication: A. S. Eddington, *Report on the Relativity Theory of Gravitation* (Minkowski Institute Press, Montreal 2014)
5. S. Chandrasekhar, *Eddington – The most distinguished astrophysicist of his time* (Cambridge University Press, Cambridge 1983), p. 24
6. E. B. Wilson, Review: A. S. Eddington, Space, Time and Gravitation; an Outline of the General Relativity Theory, *Bull. Amer. Math. Soc.*, Volume 27, Number 4 (1921), 182-186
7. A. S. Eddington, *The Mathematical Theory of Relativity* (Cambridge University Press, 1923); new publication: A. S. Eddington, *The Mathematical Theory of Relativity* (Minkowski Institute Press, Montreal 2016)
8. H. Minkowski, Space and Time. In: H. Minkowski, *Space and Time: Minkowski's papers on relativity* (Minkowski Institute Press, Montreal 2012)

Preface

By his theory of relativity Albert Einstein has provoked a revolution of thought in physical science.

The achievement consists essentially in this:—Einstein has succeeded in separating far more completely than hitherto the share of the observer and the share of external nature in the things we see happen. The perception of an object by an observer depends on his own situation and circumstances; for example, distance will make it appear smaller and dimmer. We make allowance for this almost unconsciously in interpreting what we see. But it now appears that the allowance made for the *motion* of the observer has hitherto been too crude—a fact overlooked because in practice all observers share nearly the same motion, that of the earth. Physical space and time are found to be closely bound up with this motion of the observer; and only an amorphous combination of the two is left inherent in the external world. When space and time are relegated to their proper source—the observer—the world of nature which remains appears strangely unfamiliar; but it is in reality simplified, and the underlying unity of the principal phenomena is now clearly revealed. The deductions from this new outlook have, with one doubtful exception, been confirmed when tested by experiment.

It is my aim to give an account of this work without introducing anything very technical in the way of mathematics, physics, or philosophy. The new view of space and time, so opposed to our habits of thought, must in any case demand unusual mental exercise. The results appear strange; and the incongruity is not without a humorous side. For the first nine chapters the task is one of interpreting a clear-cut theory, accepted in all its essentials by a large and growing school of physicists—although perhaps not everyone would accept the author's views of its meaning. Chap. X and Chap. XI deal with very recent advances, with regard to which opinion is more fluid. As for the last chapter, containing the author's speculations on the meaning of nature, since it touches on the rudiments of a philosophical system, it is perhaps too sanguine to hope that it can ever be other than controversial.

A non-mathematical presentation has necessary limitations; and the reader who wishes to learn how certain exact results follow from Einstein's, or even Newton's, law of gravitation is bound to seek the reasons in a mathematical treatise. But this limitation of range is perhaps less serious than the limitation of intrinsic truth. There is a relativity of truth, as there is a relativity of space.—

> "For IS and IS-NOT though *with* Rule and Line
> And UP-AND-DOWN *without*, I could define."

Alas! It is not so simple. We abstract from the phenomena that which is peculiar to the position and motion of the observer; but can we abstract that which is peculiar to the limited imagination of the human brain? We think we can, but only in the symbolism of mathematics. As the language of a poet rings with a truth that eludes the clumsy explanations of his commentators, so the geometry of relativity in its perfect harmony expresses a truth of form and type in nature, which my bowdlerised version misses.

But the mind is not content to leave scientific Truth in a dry husk of mathematical symbols, and demands that it shall be alloyed with familiar images. The mathematician, who handles x so lightly, may fairly be asked to state, not indeed the inscrutable meaning of x in nature, but the meaning which x conveys to *him*.

Although primarily designed for readers without technical knowledge of the subject, it is hoped that the book may also appeal to those who have gone into the subject more deeply. A few notes have been added in the Appendix mainly to bridge the gap between this and more mathematical treatises, and to indicate the points of contact between the argument in the text and the parallel analytical investigation.

It is impossible adequately to express my debt to contemporary literature and discussion. The writings of Einstein, Minkowski, Hilbert, Lorentz, Weyl, Robb, and others, have provided the groundwork; in the give and take of debate with friends and correspondents, the extensive ramifications have gradually appeared.

1 May, 1920 A. S. E.

Contents

Prologue
What is Geometry?

A conversation between—
An experimental Physicist.
A pure Mathematician.
A Relativist, who advocates the newer conceptions of time and space in physics.

Rel. There is a well-known proposition of Euclid which states that "Any two sides of a triangle are together greater than the third side." Can either of you tell me whether nowadays there is good reason to believe that this proposition is true?

Math. For my part, I am quite unable to say whether the proposition is true or not. I can deduce it by trustworthy reasoning from certain other propositions or axioms, which are supposed to be still more elementary. If these axioms are true, the proposition is true; if the axioms are not true, the proposition is not true universally. Whether the axioms are true or not I cannot say, and it is outside my province to consider.

Phys. But is it not claimed that the truth of these axioms is self-evident?

Math. They are by no means self-evident to me; and I think the claim has been generally abandoned.

Phys. Yet since on these axioms you have been able to found a logical and self-consistent system of geometry, is not this indirect evidence that they are true?

Math. No. Euclid's geometry is not the only self-consistent system of geometry. By choosing a different set of axioms I can, for example, arrive at Lobatchewsky's geometry, in which many of the propositions of Euclid are not in general true. From my point of view there is nothing to choose between these different geometries.

Rel. How is it then that Euclid's geometry is so much the most important system?

Math. I am scarcely prepared to admit that it is the most important. But for reasons which I do not profess to understand, my friend the Physicist is more interested in Euclidean geometry than in any other, and is continually setting us problems in it. Consequently we have tended to give an undue share of attention to the Euclidean system. There have, however, been great geometers like Riemann who have done something to restore a proper perspective.

Rel. (to Physicist). Why are you specially interested in Euclidean geometry? Do you believe it to be the true geometry?

Phys. Yes. Our experimental work proves it true.

Rel. How, for example, do you prove that any two sides of a triangle are together greater than the third side?

Phys. I can, of course, only prove it by taking a very large number of typical cases, and I am limited by the inevitable inaccuracies of experiment. My proofs are not so general or so perfect as those of the pure mathematician. But it is a recognised principle in physical science that it is permissible to generalise from a reasonably wide range of experiment; and this kind of proof satisfies me.

Rel. It will satisfy me also. I need only trouble you with a special case. Here is a triangle ABC; how will you prove that $AB + BC$ is greater than AC?

Phys. I shall take a scale and measure the three sides.

Rel. But we seem to be talking about different things. I was speaking of a proposition of geometry – properties of space, not of matter. Your experimental proof only shows how a material scale behaves when you turn it into different positions.

Phys. I might arrange to make the measures with an optical device.

Rel. That is worse and worse. Now you are speaking of properties of light.

Phys. I really cannot tell you anything about it, if you will not let me make measurements of any kind. Measurement is my only means of finding out about nature. I am not a metaphysicist.

Rel. Let us then agree that by *length* and *distance* you always mean a quantity arrived at by measurements with material or optical appliances. You have studied experimentally the laws obeyed by these *measured lengths*, and have found the geometry to which they conform. We will call this geometry "Natural Geometry"; and it evidently has much greater importance for you than any other of the systems which the brain of the mathematician has invented. But we must remember that its subject matter involves the behaviour of material scales – the properties of matter. Its laws are just as much laws of physics as, for example, the laws of electromagnetism.

Phys. Do you mean to compare space to a kind of magnetic field? I scarcely understand.

Rel. You say that you cannot explore the world without some kind of apparatus. If you explore with a scale, you find out the natural geometry; if you explore with a magnetic needle, you find out the magnetic field. What we may call the field of extension, or space-field, is just as much a physical quality as the magnetic field. You can think of them both existing together in the aether, if you like. The laws of both must be determined by experiment. Of course, certain approximate laws of the space-field (Euclidean geometry) have been familiar to us from childhood; but we must get rid of the idea that there is anything inevitable about these laws, and that it would be impossible to find in other parts of the universe space-fields where these laws do not apply. As to how far space really resembles a magnetic field, I do not wish to dogmatise; my point is that they present themselves to experimental investigation in very much the same way.

Let us proceed to examine the laws of natural geometry. I have a tape-measure, and here is the triangle. $AB = 39\frac{1}{2}$ in., $BC = \frac{1}{8}$ in., $CA = 39\frac{7}{8}$ in. Why, your proposition does not hold!

Phys. You know very well what is wrong. You gave the tape-measure a big stretch when you measured AB.

Rel. Why shouldn't I?

Phys. Of course, a length must be measured with a rigid scale.

Rel. That is an important addition to our definition of length. But what is a rigid scale?

Phys. A scale which always keeps the same length.

Rel. But we have just defined length as the quantity arrived at by measures with a rigid scale; so you will want another rigid scale to test whether the first one changes length; and a third to test the second; and so *ad infinitum.* You remind me of the incident of the clock and time-gun in Egypt. The man in charge of the time-gun fired it by the clock; and the man in charge of the clock set it right by the time-gun. No, you must not define length by means of a rigid scale, and define a rigid scale by means of length.

Phys. I admit I am hazy about strict definitions. There is not time for everything; and there are so many interesting things to find out in physics, which take up my attention. Are you so sure that you are prepared with a logical definition of all the terms you use?

Rel. Heaven forbid! I am not naturally inclined to be rigorous about these things. Although I appreciate the value of the work of those who are digging at the foundations of science, my own interests are mainly in the upper structure. But sometimes, if we wish to add another storey, it is necessary to deepen the foundations. I have a definite object in trying to arrive at the exact meaning of length. A strange theory is floating round, to which you may feel initial objections; and you probably would not wish to let your views go by default. And after all, when you claim to determine lengths to eight significant figures, you must have a pretty definite standard of right and wrong measurements.

Phys. It is difficult to define what we mean by rigid; but in practice we can tell if a scale is likely to change length appreciably in different circumstances.

Rel. No. Do not bring in the idea of change of length in describing the apparatus for defining length. Obviously the adopted standard of length cannot change length, whatever it is made of. If a metre is defined as the length of a certain bar, that bar can never be anything but a metre long; and if we assert that this bar changes length, it is clear that we must have changed our minds as to the definition of length. You recognised that my tape-measure was a defective standard– that it was not rigid. That was not because it changed length, because, if it was the standard of length, it could not change length. It was lacking in some other quality.

You know an approximately rigid scale when you see one. What you are comparing it with is not some non-measurable ideal of length, but some attainable, or at least approachable, ideal of material constitution. Ordinary scales have defects – flexure, expansion with temperature, etc. – which can be reduced by suitable precautions; and the limit, to which you approach as you reduce them, is your rigid scale. You can define these defects without appealing to any extraneous definition of length; for example, if you have two rods of the same material whose extremities are just in contact with one another, and when one of them is heated the extremities no longer can be adjusted to coincide, then the material has a temperature-coefficient of expansion. Thus you can compare experimentally the temperature-coefficients of different metals and arrange them in diminishing sequence. In this sort of way you can specify the nature of your ideal rigid rod, before you introduce the term length.

Phys. No doubt that is the way it should be defined.

Rel. We must recognise then that all our knowledge of space rests on the behaviour of material measuring-scales free from certain definable defects of constitution.

Phys. I am not sure that I agree. Surely there is a sense in which the statement $AB = 2CD$ is true or false, even if we had no conception of a material measuring-rod. For instance, there is, so to speak, twice as much paper between A and B, as between C and D.

Rel. Provided the paper is uniform. But then, what does uniformity of the paper mean? That the amount in given length is constant. We come back at once to the need of defining length.

If you say instead that the amount of "space" between A and B is twice that between C and D, the same thing applies. You imagine the intervals filled with uniform space; but the uniformity simply means that the same amount of space corresponds to each inch of your rigid measuring-rod. You have arbitrarily used your rod to divide space into so-called equal lumps. It all comes back to the rigid rod.

I think you were right at first when you said that you could not find out anything without measurement; and measurement involves some specified material appliance.

Now you admit that your measures cannot go beyond a certain close approximation, and that you have not tried all possible conditions. Supposing that one corner of your triangle was in a very intense gravitational field – far stronger than any we have had experience of – I have good ground for believing that under those conditions you might find the sum of two sides of a triangle, as measured with a rigid rod, appreciably less than the third side. In that case would you be prepared to give up Euclidean geometry?

Phys. I think it would be risky to assume that the strong force of gravitation made no difference to the experiment.

Rel. On my supposition it makes an important difference.

Phys. I mean that we might have to make corrections to the measures, because the action of the strong force might possibly distort the measuring-rod.

Rel. In a rigid rod we have eliminated any special response to strain.

Phys. But this is rather different. The extension of the rod is determined by the positions taken up by the molecules under the forces to which they are subjected; and there might be a response to the gravitational force which all kinds of matter would share. This could scarcely be regarded as a defect; and our so-called rigid rod would not be free from it any more than any other kind of matter.

Rel. True; but what do you expect to obtain by correcting the measures? You correct measures, when they are untrue to standard. Thus you correct the readings of a hydrogen-thermometer to obtain the readings of a perfect gas-thermometer, because the hydrogen molecules have finite size, and exert special attractions on one another, and you prefer to take as standard an ideal gas with infinitely small molecules. But in the present case, what is the standard you are aiming at when you propose to correct measures made with the rigid rod?

Phys. I see the difficulty. I have no knowledge of space apart from my measures, and I have no better standard than the rigid rod. So it is difficult to see what the corrected measures would mean. And yet it would seem to me more natural to suppose that the failure of the proposition was due to the measures going wrong rather than to an alteration in the character of space.

Rel. Is not that because you are still a bit of a metaphysicist? You keep some notion of a space which is superior to measurement, and are ready to throw over the measures rather than let this space be distorted. Even if there were reason for believing in such a space, what possible reason could there be for assuming it to be Euclidean? Your sole reason for believing space to be Euclidean is that hitherto your measures have made it appear so; if now measures of certain parts of space prefer non-Euclidean geometry, all reason for assuming Euclidean space disappears. Mathematically and conceptually Euclidean and non-Euclidean space are on the same footing; our preference for Euclidean space was based on measures, and must stand or fall by measures.

Phys. Let me put it this way. I believe that I am trying to measure something called length, which has an absolute meaning in nature, and is of importance in connection with the laws of nature. This length obeys Euclidean geometry. I believe my measures with a rigid rod determine it accurately when no disturbance like gravitation is present; but in a gravitational field it is not unreasonable to expect that the uncorrected measures may not give it exactly.

Rel. You have three hypotheses there: – (1) there is an absolute thing in nature corresponding to length, (2) the geometry of these absolute lengths is Euclidean, and (3) practical measures determine this length accurately when there is no gravitational force. I see no necessity for these hypotheses, and propose to do without them. *Hypotheses non fingo.* The second hypothesis seems to me particularly objectionable. You assume that this absolute thing in nature obeys the laws of Euclidean geometry. Surely it is contrary to scientific principles to lay down arbitrary laws for nature to obey; we must find out her laws by experiment. In this case the only experimental evidence is that measured lengths (which by your own admission are not necessarily the same as this absolute thing) sometimes obey Euclidean geometry and sometimes do not. Again it would seem reasonable to doubt your third hypothesis beyond, say, the sixth decimal place; and that would play havoc with your more delicate measures. But where I fundamentally differ from you is the first hypothesis. Is there some absolute quantity in nature that we try to determine when we measure length? When we try to determine the number of molecules in a given piece of matter, we have to use indirect methods, and different methods may give systematically different results; but no one doubts that there is a definite number of molecules, so that there is some meaning in saying that certain methods are theoretically good and others inaccurate. Counting appears to be an absolute operation. But it seems to me that other physical measures are on a different footing. Any physical quantity, such as length, mass, force, etc., which is not a pure number, can only be defined as the result arrived at by conducting a physical experiment according to specified rules.

So I cannot conceive of any "length" in nature independent of a definition of the way of measuring length. And, if there is, we may disregard it in physics, because it is beyond the range of experiment. Of course, it is always possible that we may come across some quantity, not given directly by experiment, which plays a fundamental part in theory. If so, it will turn up in due course in our theoretical formulae. But it is no good assuming such a quantity, and laying down *a priori* laws for it to obey, on the off-chance of its proving useful.

Phys. Then you will not let me blame the measuring-rod when the proposition fails?

Rel. By all means put the responsibility on the measuring-rod. Natural geometry is the theory of the behaviour of material scales. Any proposition in natural geometry is an assertion as to the behaviour of rigid scales, which must accordingly take the blame or credit. But do not say that the rigid scale is wrong, because that implies a standard of right which does not exist.

Phys. The space which you are speaking of must be a sort of abstraction of the extensional relations of matter.

Rel. Exactly so. And when I ask you to believe that space can be non-Euclidean, or, in popular phrase, warped, I am not asking you for any violent effort of the imagination; I only mean that the extensional relations of matter obey somewhat modified laws. Whenever we investigate the properties of space experimentally, it is these extensional relations that we are finding. Therefore it seems logical to conclude that space as known to us must be the abstraction of these material relations, and not something more transcendental. The reformed methods of teaching geometry in schools would be utterly condemned, and it would be misleading to set schoolboys to verify propositions of geometry by measurement, if the space they are supposed to be studying had not this meaning.

I suspect that you are doubtful whether this abstraction of extensional relations quite fulfils your general idea of space; and, as a necessity of thought, you require something beyond. I do not think I need disturb that impression, provided you realise that it is not the properties of this more transcendental thing we are speaking of when we describe geometry as Euclidean or non-Euclidean.

Math. The view has been widely held that space is neither physical nor metaphysical, but conventional. Here is a passage from Poincaré's *Science and Hypothesis*, which describes this alternative idea of space:

"If Lobatchewsky's geometry is true, the parallax of a very distant star will be finite. If Riemann's is true, it will be negative. These are the results which seem within the reach of experiment, and it is hoped that astronomical observations may enable us to decide between the two geometries. But what we call a straight line in astronomy is simply the path of a ray of light. If, therefore, we were to discover negative parallaxes, or to prove that all parallaxes are higher than a certain limit, we should have a choice between two conclusions: we could give up Euclidean geometry, or modify the laws of optics, and suppose that light is not rigorously propagated in a straight line. It is needless to add that everyone would look upon this solution as the more advantageous. Euclidean geometry, therefore, has nothing to fear from fresh experiments."

Rel. Poincaré's brilliant exposition is a great help in understanding the problem now confronting us. He brings out the interdependence between geometrical laws and physical laws, which we have to bear in mind continually. We can add on to one set of laws that which we subtract from the other set. I admit that space is conventional – for that matter, the meaning of every word in the language is conventional. Moreover, we have actually arrived at the parting of the ways imagined by Poincaré, though the crucial experiment is not precisely the one he mentions. But I deliberately adopt the alternative, which, he takes for granted, everyone would consider less advantageous. I call the space thus chosen *physical space*, and its geometry *natural geometry*, thus admitting that other conventional meanings of space and geometry are possible. If it were only a question of the meaning of space – a rather vague term – these other possibilities might have some advantages. But the meaning assigned to length and distance has to go along with the meaning assigned to space. Now these are quantities which the physicist has been accustomed to measure with great accuracy; and they enter fundamentally into the whole of our experimental knowledge of the world. We have a knowledge of the so-called extent of the stellar universe, which, whatever it may amount to in terms of ultimate reality, is not a mere description of location in a conventional and arbitrary mathematical space. Are we to be robbed of the terms in which we are accustomed to describe that knowledge?

The law of Boyle states that the pressure of a gas is proportional to its density. It is found by experiment that this law is only approximately true. A certain mathematical simplicity would be gained by conventionally redefining *pressure* in such a way that Boyle's law would be rigorously obeyed. But it would be high-handed to appropriate the word pressure in this way, unless it had been ascertained that the physicist had no further use for it in its original meaning.

Phys. I have one other objection. Apart from measures, we have a general perception of space, and the space we perceive is at least approximately Euclidean.

Rel. Our perceptions are crude measures. It is true that our perception of space is very largely a matter of optical measures with the eyes. If in a strong gravitational field optical and mechanical measures diverged, we should have to make up our minds which was the preferable standard, and afterwards abide by it. So far as we can ascertain, however, they agree in all circumstances, and no such difficulty arises. So, if physical measures give us a non-Euclidean space, the space of perception will be non-Euclidean. If you were transplanted into an extremely intense gravitational field, you would directly perceive the non-Euclidean properties of space.

Phys. Non-Euclidean space seems contrary to reason.

Math. It is not contrary to reason, but contrary to common experience, which is a very different thing, since experience is very limited.

Phys. I cannot imagine myself perceiving non-Euclidean space!

Math. Look at the reflection of the room in a polished doorknob, and imagine yourself one of the actors in what you see going on there.

Rel. I have another point to raise. The distance between two points is to be the length measured with a rigid scale. Let us mark the two points by particles of matter, because we must somehow identify them by reference to material objects. For simplicity we shall suppose that the two particles have no relative motion, so that the distance – whatever it is – remains constant. Now you will probably agree that there is no such thing as absolute motion; consequently there is no standard condition of the scale which we can call "at rest." We may measure with the scale moving in any way we choose, and if results for different motions disagree, there is no criterion for selecting the true one. Further, if the particles are sliding past the scale, it makes all the difference what instants we choose for making the two readings.

Phys. You can avoid that by defining distance as the measurement made with a scale which has the same velocity as the two points. Then they will always be in contact with two particular divisions of the scale.

Rel. A very sound definition; but unfortunately it does not agree with the meaning of distance in general use. When the relativist wishes to refer to this length, he calls it the *proper-length*; in non-relativity physics it does not seem to have been used at all. You see it is not convenient to send your apparatus hurling through the laboratory – after a pair of α particles, for example. And you could scarcely measure the length of a wave of light by this convention[1]. So the physicist refers his lengths to apparatus at rest on the earth; and the mathematician starts with the words "Choose unaccelerated rectangular axes Ox, Oy, Oz, ..." and assumes that the measuring-scales are at rest relatively to these axes. So when the term length is used some arbitrary standard motion of the measuring apparatus must always be implied.

Phys. Then if you have fixed your standard motion of the measuring-rod, there will be no ambiguity if you take the readings of both particles at the same moment.

Rel. What is the same moment at different places? The conception of simultaneity in different places is a difficult one. Is there a particular instant in the progress of time on another world, Arcturus, which is the same as the present instant on the Earth?

Phys. I think so, if there is any connecting link. We can observe an event, say a change of brightness, on Arcturus, and, allowing for the time taken by light to travel the distance, determine the corresponding instant on the earth.

Rel. But then you must know the speed of the earth through the aether. It may have shortened the light-time by going some way to meet the light coming from Arcturus.

Phys. Is not that a small matter?

Rel. At a very modest reckoning the motion of the earth in the interval might alter the light-time by several days. Actually, however, any speed of the earth through the aether up to the velocity of light is admissible, without affecting anything observable. At least, nothing has been discovered which contradicts this. So the error may be months or years.

Phys. What you have shown is that we have not sufficient knowledge to determine in practice which are simultaneous events on the Earth and Arcturus. It does not follow that there is no definite simultaneity.

Rel. That is true, but it is at least possible that the reason why we are unable to determine simultaneity in practice (or, what comes to pretty much the same thing, our motion through the aether) in spite of many brilliant attempts, is that there is no such thing as absolute simultaneity of distant events. It is better therefore not to base our physics on this notion of absolute simultaneity, which may turn out not to exist, and is in any case out of reach at present.

[1] The proper-length of a light-wave is actually infinite.

But what all this comes to is that time as well as space is implied in all our measures. The fundamental measurement is not the interval between two points of space, but between two points of space associated with instants of time.

Our natural geometry is incomplete at present. We must supplement it by bringing in time as well as space. We shall need a perfect clock as well as a rigid scale for our measures. It may be difficult to choose an ideal standard clock; but whatever definition we decide on must be a physical definition. We must not dodge it by saying that a perfect clock is one which keeps perfect time. Perhaps the best theoretical clock would be a pulse of light travelling in vacuum to and fro between mirrors at the ends of a rigid scale. The instants of arrival at one end would define equal intervals of time.

Phys. I think your unit of time would change according to the motion of your "clock" through the aether.

Rel. Then you are comparing it with some notion of absolute time. I have no notion of time except as the result of measurement with some kind of clock. (Our immediate perception of the flight of time is presumably associated with molecular processes in the brain which play the part of a material clock.) If you know a better clock, let us adopt it; but, having once fixed on our ideal clock there can be no appeal from its judgments. You must remember too that if you wish to measure a second *at one place*, you must keep your clock fixed at what you consider to be one place; so its motion is defined. The necessity of defining the motion of the clock emphasises that one cannot consider time apart from space; there is one geometry comprising both.

Phys. Is it right to call this study *geometry*? Geometry deals with space alone.

Math. I have no objection. It is only necessary to consider time as a fourth dimension. Your complete natural geometry will be a geometry of four dimensions.

Phys. Have we then found the long-sought fourth dimension?

Math. It depends what kind of a fourth dimension you were seeking. Probably not in the sense you intend. For me it only means adding a fourth variable, t, to my three space-variables x, y, z. It is no concern of mine what these variables really represent. You give me a few fundamental laws that they satisfy, and I proceed to deduce other consequences that may be of interest to you. The four variables may for all I know be the pressure, density, temperature and entropy of a gas; that is of no importance to me. But you would not say that a gas had four dimensions because four mathematical variables were used to describe it. Your use of the term "dimensions" is probably more restricted than mine.

Phys. I know that it is often a help to represent pressure and volume as height and width on paper; and so geometry may have applications to the theory of gases. But is it not going rather far to say that geometry can deal directly with these things and is not necessarily concerned with lengths in space?

Math. No. Geometry is nowadays largely analytical, so that in form as well as in effect, it deals with variables of an unknown nature. It is true that I can often see results more easily by taking my x and y as lengths on a sheet of paper. Perhaps it would be helpful in seeing other results if I took them as pressure and density in a steam-engine; but a steam-engine is not so handy as a pencil. It is literally true that I do not want to know the significance of the variables x, y, z, t that I am discussing. That is lucky for the Relativist, because although he has defined carefully how they are to be measured, he has certainly not conveyed to me any notion of how I am to picture them, if my picture of absolute space is an illusion.

Phys. Yours is a strange subject. You told us at the beginning that you are not concerned as to whether your propositions are true, and now you tell us you do not even care to know what you are talking about.

Math. That is an excellent description of Pure Mathematics, which has already been given by an eminent mathematician[2].

Rel. I think there is a real sense in which time is a fourth dimension – as distinct from a fourth variable. The term dimension seems to be associated with relations of *order*. I believe that the order of events in nature is one indissoluble four-dimensional order. We may split it arbitrarily into space and time, just as we can split the order of space into length, breadth and thickness. But space without time is as incomplete as a surface without thickness.

Math. Do you argue that the real world behind the phenomena is four-dimensional?

Rel. I think that in the real world there must be a set of entities related to one another in a four-dimensional order, and that these are the basis of the perceptual world so far as it is yet explored by physics. But it is possible to pick out a four-dimensional set of entities from a basal world of five dimensions, or even of three dimensions. The straight lines in three-dimensional space form a four-dimensional set of entities, i.e. they have a four-fold order. So one cannot predict the ultimate number of dimensions in the world – if indeed the expression *dimensions* is applicable.

Phys. What would a philosopher think of these conceptions? Or is he solely concerned with a metaphysical space and time which is not within reach of measurement.

Rel. In so far as he is a psychologist our results must concern him. Perception is a kind of crude physical measurement; and perceptual space and time is the same as the measured space and time, which is the subject-matter of natural geometry. In other respects he may not be so immediately concerned. Physicists and philosophers have long agreed that motion through absolute space can have no meaning; but in physics the question is whether motion through aether has any meaning. I consider that it has no meaning; but that answer, though it brings philosophy and physics into closer relation, has no bearing on the philosophic question of absolute motion. I think, however, we are entitled to expect a benevolent interest from philosophers, in that we are giving to their ideas a perhaps unexpected practical application.

Let me now try to sum up my conclusions from this conversation. We have

[2]"Pure mathematics consists entirely of such asseverations as that, if such and such a proposition is true of *anything*, then such and such a proposition is true of that thing. It is essential not to discuss whether the first proposition is really true, and not to mention what the anything is of which it is supposed to be true.... Thus mathematics may be defined as the subject in which we never know what we are talking about, nor whether what we are saying is true." BERTRAND RUSSELL.

been trying to give a precise meaning to the term *space*, so that we may be able to determine exactly the properties of the space we live in. There is no means of determining the properties of our space by *a priori* reasoning, because there are many possible kinds of space to choose from, no one of which can be considered more likely than any other. For more than 2000 years we have believed in a Euclidean space, because certain experiments favoured it; but there is now reason to believe that these same experiments when pushed to greater accuracy decide in favour of a slightly different space (in the neighbourhood of massive bodies). The relativist sees no reason to change the rules of the game because the result does not agree with previous anticipations. Accordingly when he speaks of space, he means the space revealed by measurement, whatever its geometry. He points out that this is the space with which physics is concerned; and, moreover, it is the space of everyday perception. If his right to appropriate the term space in this way is challenged, he would urge that this is the sense in which the term has always been used in physics hitherto; it is only recently that conservative physicists, frightened by the revolutionary consequences of modern experiments, have begun to play with the idea of a pre-existing space whose properties cannot be ascertained by experiment – a metaphysical space, to which they arbitrarily assign Euclidean properties, although it is obvious that its geometry can never be ascertained by experiment. But the relativist, in defining space as *measured space*, clearly recognises that all measurement involves the use of material apparatus; the resulting geometry is specifically a study of the extensional relations of matter. He declines to consider anything more transcendental.

My second point is that since natural geometry is the study of extensional relations of natural objects, and since it is found that their space-order cannot be discussed without reference to their time-order as well, it has become necessary to extend our geometry to four dimensions in order to include time.

1 THE FITZGERALD CONTRACTION

In order to reach the Truth, it is necessary, once in one's life, to put everything in doubt—so far as possible.

DESCARTES.

WILL it take longer to swim to a point 100 yards up-stream and back, or to a point 100 yards across-stream and back?

In the first case there is a long toil up against the current, and then a quick return helped by the current, which is all too short to compensate. In the second case the current also hinders, because part of the effort is devoted to overcoming the drift down-stream. But no swimmer will hesitate to say that the hindrance is the greater in the first case.

Let us take a numerical example. Suppose the swimmer's speed is 50 yards a minute in still water, and the current is 30 yards a minute. Thus the speed against the current is 20, and with the current 80 yards a minute. The up journey then takes 5 minutes and the down journey $1\frac{1}{4}$ minutes. Total time, $6\frac{1}{4}$ minutes.

Going across-stream the swimmer must aim at a point E above the point B where he wishes to arrive, so that OE represents his distance travelled in still water, and EB the amount he has drifted down. These must be in the ratio 50 to 30, and we then know from the right-angled triangle OBE that OB will correspond to 40. Since OB is 100 yards, OE is 125 yards, and the time taken is $2\frac{1}{2}$ minutes. Another $2\frac{1}{2}$ minutes will be needed for the return journey. Total time, 5 minutes.

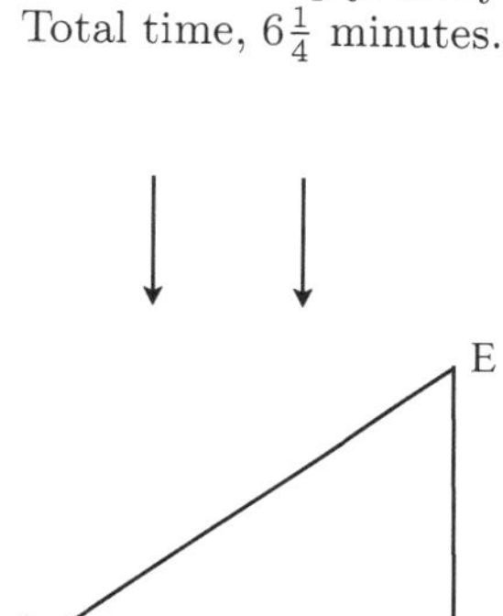

FIG. 1

In still water the time would have been 4 minutes.

The up-and-down swim is thus longer than the transverse swim in the ratio $6\frac{1}{4} : 5$ minutes. Or we may write the ratio

$$\frac{1}{\sqrt{(1-(\frac{30}{50})^2)}}$$

which shows how the result depends on the ratio of the speed of the current to the speed of the swimmer, viz. $\frac{30}{50}$.

A very famous experiment on these lines was tried in America in the year 1887. The swimmer was a wave of light, which we know swims through the aether with a speed of 186,330 miles a second. The aether was flowing through the laboratory like a river past its banks. The light-wave was divided, by partial reflection at a thinly silvered surface, into two parts, one of which was set to perform the up-and-down stream journey and the other the across-stream journey. When the two waves reached their proper turning-points they were sent back to the starting-point by mirrors. To judge the result of the race, there was an optical device for studying interference fringes; because the recomposition of the two waves after the journey would reveal if one had been delayed more than the other, so that, for example, the crest of one instead of fitting on to the crest of the other coincided with its trough.

To the surprise of Michelson and Morley, who conducted the experiment, the result was a dead-heat. It is true that the direction of the current of aether was not known—they hoped to find it out by the experiment. That, however, was got over by trying a number of different orientations. Also it was possible that there might actually be no current at a particular moment. But the earth has a velocity of $18\frac{1}{2}$ miles a second, continually changing direction as it goes round the sun; so that at some time during the year the motion of a terrestrial laboratory through the aether must be at least $18\frac{1}{2}$ miles a second. The experiment should have detected the delay by a much smaller current; in a repetition of it by Morley and Miller in 1905, a current of 2 miles a second would have been sufficient.

If we have two competitors, one of whom is known to be slower than the other, and yet they both arrive at the winning-post at the same time, it is clear that they cannot have travelled equal courses. To test this, the whole apparatus was rotated through a right angle, so that what had been the up-and-down course became the transverse course, and *vice versa*. Our two competitors interchanged courses, but still the result was a dead-heat.

The surprising character of this result can be appreciated by contrasting it with a similar experiment on sound-waves. Sound consists of waves in air or other material, as light consists of waves in aether. It would be possible to make a precisely similar experiment on sound, with a current of air past the apparatus instead of a current of aether. In that case the greater delay of the wave along the direction of the current would certainly show itself experimentally. Why does light seem to behave differently?

The straightforward interpretation of this remarkable result is that each course undergoes an automatic contraction when it is swung from the transverse to the longitudinal position, so that whichever arm of the apparatus is placed up-stream it straightway becomes the shorter. The course is marked out in the rigid material apparatus, and we have to suppose that the length of any part of the apparatus changes as it is turned in different directions with respect to the aether-current. It is found that the kind of material—metal, stone or wood—makes no difference to the experiment. The contraction must be the same for all kinds of matter; the expected delay depends only on the ratio of the speed of the aether current to the speed of light, and the contraction which compensates it must be equally definite.

This explanation was proposed by FitzGerald, and at first sight it seems a strange and arbitrary hypothesis. But it has been rendered very plausible by subsequent theoretical researches of Larmor and Lorentz. Under ordinary circumstances the form and size of a solid body is maintained by the forces of cohesion between its particles. What is the nature of cohesion? We guess that it is made up of electric forces between the molecules. But the aether is the medium in which electric force has its seat; hence it will not be a matter of indifference to these forces how the electric medium is flowing with respect to the molecules. When the flow changes there will be a readjustment of cohesive forces, and we must expect the body to take a new shape and size.

The theory of Larmor and Lorentz enables us to trace in detail the readjustment. Taking the accepted formulae of electromagnetic theory, they showed that the new form of equilibrium would be contracted in just such a way and by just such an amount as FitzGerald's explanation requires[1].

The contraction in most cases is extremely minute. We have seen that when the ratio of the speed of the current to that of the swimmer is $\frac{3}{5}$, a contraction in the ratio $\sqrt{(1-(\frac{3}{5})^2)}$ is needed to compensate for the delay. The earth's orbital velocity is $\frac{1}{10000}$ of the velocity of light, so that it will give a contraction of $\sqrt{(1-(\frac{1}{10000})^2)}$, or 1 part in $200,000,000$. This would mean that the earth's diameter in the direction of its motion is shortened by $2\frac{1}{2}$ inches.

The Michelson-Morley experiment has thus failed to detect our motion through the aether, because the effect looked for—the delay of one of the light waves—is exactly compensated by an automatic contraction of the matter forming the apparatus. Other ingenious experiments have been tried, electrical and optical experiments of a more technical nature. They likewise have failed, because there is always an automatic compensation somewhere. We now believe there is something in the nature of things which inevitably makes these compensations, so that it will never be possible to determine our motion through the aether. Whether we are at rest in it, or whether we are rushing through it with a speed not much less than that of light, will make no difference to anything that can possibly be observed.

This may seem a rash generalization from the few experiments actually performed; more particularly, since we can only experiment with the small range of velocity caused by the earth's orbital motion. With a larger range residual differences might be disclosed. But there is another reason for believing that the compensation is not merely approximate but exact. The compensation has been traced theoretically to its source in the well-known laws of electromagnetic force; and here it is mathematically exact. Thus the generalization is justified, at least in so far as the observed phenomena depend on electromagnetic causes, and in so far as the universally accepted laws of electromagnetism are accurate.

The generalization here laid down is called the restricted Principle of Relativity: *It is impossible by any experiment to detect uniform motion relative to the aether.*

There are other natural forces which have not as yet been recognised as coming within the electromagnetic scheme—gravitation, for example—and for these other tests are required. Indeed we were scarcely justified in stating above that the diameter of the earth would contract $2\frac{1}{2}$ inches, because the figure of the earth is determined mainly by gravitation, whereas the Michelson-Morley experiment relates to bodies held together by cohesion. There is fair evidence of a rather technical kind that the compensation exists also for phenomena in which gravitation is concerned; and we shall assume that the principle covers all

[1] Appendix, Note 1.

the forces of nature.

Suppose for a moment it were not so, and that it were possible to determine a kind of absolute motion of the earth by experiments or observations involving gravitation. Would this throw light on our motion through the aether? I think not. It would show that there is some standard of rest with respect to which the law of gravitation takes a symmetrical and simple form; presumably this standard corresponds to some gravitational medium, and the motion determined would be motion with respect to that medium. Similarly if the motion were revealed by vital or psychical phenomena, it would be motion relative to some vital or psychical medium. The aether, defined as the seat of electric forces, must be revealed, if at all, by electric phenomena.

It is well to remember that there is reasonable justification for adopting the principle of relativity even if the evidence is insufficient to prove it. In Newtonian dynamics the phenomena are independent of uniform motion of the system; no explanation is asked for, because it is difficult to see any reason why there should be an effect. If in other phenomena the principle fails, then we must seek for an explanation of its failure—and no doubt a plausible explanation can be devised; but so long as experiment gives no indication of a failure, it is idle to anticipate such a complication. Clearly physics cannot concern itself with all the possible complexities which *may* exist in nature, but have not hitherto betrayed themselves in any experiment.

The principle of relativity has implications of a most revolutionary kind. Let us consider what is perhaps an exaggerated case—or perhaps the actual case, for we cannot tell. Let the reader suppose that he is travelling through the aether at 161,000 miles a second vertically upwards; if he likes to make the positive assertion that this is his velocity, no one will be able to find any evidence to contradict him. For this speed the FitzGerald contraction is just $\frac{1}{2}$, so that every object contracts to half its original length when turned into the vertical position.

As you lie in bed, you are, say, 6 feet long. Now stand upright; you are 3 feet. You are incredulous? Well, let us prove it! Take a yard-measure; when turned vertically it must undergo the FitzGerald contraction, and become only half a yard. If you measure yourself with it, you will find you are just two—*half-yards*. "But I can see that the yard-measure does not change length when I turn it." What you perceive is an image of the rod on the retina of your eye; you imagine that the image occupies the same space in both positions; but your retina has contracted in the vertical direction without your knowing it, so that your visual estimates of vertical length are double what they should be. And so on with every test you can devise. Because everything is altered in the same way, nothing appears to be altered at all.

It is possible to devise electrical and optical tests; in that case the argument is more complicated, because we must consider the effect of the rapid current of aether on the electric forces and on waves of light. But the final conclusion is always the same; the tests will reveal nothing. Here is one illustration. To avoid distortion of the retina, lie on your back on the floor, and watch in a suitably inclined mirror someone turn the rod from the horizontal to the vertical position. You will, of course, see no change of length, and it is not possible to blame the retina this time. But is the appearance in the mirror a faithful reproduction of what is actually occurring? In a plane mirror at rest the appearance is correct; the rays of light come off the mirror at the same angle as they fall on to it, like billiard balls rebounding from an elastic cushion. But if the cushion is in rapid motion the angle of the billiard-ball will be altered; and similarly the rapid motion of the mirror through the aether alters the law of reflection. Precise calculation shows that the moving mirror will distort the image, so as to conceal exactly the changes of length which occur.

The mathematician does not need to go through all the possible tests in detail; he knows that the complete compensation is inherent in the fundamental laws of nature, and so must occur in every case. So if any suggestion is made of a device for detecting these effects, he starts at once to look for the fallacy which must surely be there. Our motion through the aether may be very much less than the value here adopted, and the changes of length may be very small; but the essential point is that they escape notice, not because they are small (if they are small), but because from their very nature they are undetectable.

There is a remarkable reciprocity about the effects of motion on length, which can best be illustrated by another example. Suppose that by development in the powers of aviation, a man flies past us at the rate of 161,000 miles a second. We shall suppose that he is in a comfortable travelling conveyance in which he can move about, and act normally and that his length is in the direction of the flight. If we could catch an instantaneous glimpse as he passed, we should see a figure about three feet high, but with the breadth and girth of a normal human being. And the strange thing is that he would be sublimely unconscious of his own undignified appearance. If he looks in a mirror in his conveyance, he sees his usual proportions; this is because of the contraction of his retina, or the distortion by the moving mirror, as already explained. But when he looks down on us, he sees a strange race of men who have apparently gone through some flattening-out process; one man looks barely 10 inches across the shoulders, another standing at right angles is almost "length and breadth, without thickness." As they turn about they change appearance like the figures seen in the old-fashioned convex-mirrors. If the reader has watched a cricket-match through a pair of prismatic binoculars, he will have seen this effect exactly.

It is the reciprocity of these appearances–that each party should think the other has contracted—that is so difficult to realise. Here is a paradox beyond even the imagination of Dean Swift. Gulliver regarded the Lilliputians as a race of dwarfs; and the Lilliputians regarded Gulliver as a giant. That is natural. If the Lilliputians had appeared dwarfs to Gulliver, and Gulliver had appeared a dwarf to the Lilliputians—but no! that is too absurd for fiction, and is an idea only to be found in the sober pages of science.

This reciprocity is easily seen to be a necessary consequence of the Principle of Relativity. The aviator must detect a FitzGerald contraction of objects moving rapidly relatively to him, just as we detect the contraction of objects moving relatively to us, and as an observer at rest in the aether detects the contraction of objects moving relatively to the aether. Any other result would indicate an observable effect due to his own motion through the aether.

Which is right? Are we or the aviator? Or are both the victims of illusion?

It is not illusion in the ordinary sense, because the impressions of both would be confirmed by every physical test or scientific calculation suggested. No one knows which is right. No one will ever know, because we can never find out which, if either, is truly at rest in the aether.

It is not only in space but in time that these strange variations occur. If we observed the aviator carefully we should infer that he was unusually slow in his movements; and events in the conveyance moving with him would be similarly retarded—as though time had forgotten to go on. His cigar lasts twice as long as one of ours. I said "infer" deliberately; we should *see* a still more extravagant slowing down of time; but that is easily explained, because the aviator is rapidly increasing his distance from us and the light-impressions take longer and longer to reach us. The more moderate retardation referred to remains after we have allowed for the time of transmission of light.

But here again reciprocity comes in, because in the aviator's opinion it is we who are travelling at 161,000 miles a second past him; and when he has made all allowances, he finds that it is we who are sluggish. Our cigar lasts twice as long as his.

Let us examine more closely how the two views are to be reconciled. Suppose we both light similar cigars at the instant he passes us. At the end of 30 minutes our cigar is finished. This signal, borne on the waves of light, hurries out at the rate of 186,000 miles a second to overtake the aviator travelling at 161,000 miles a second, who has had 30 minutes start. It will take nearly 194 minutes to overtake him, giving a total time of 224 minutes after lighting the cigar. His watch like everything else about him (including his cigar) is going at half-speed; so it records only 112 minutes elapsed when our signal arrives. The aviator knows, of course, that this is not the true time when our cigar was finished, and that he must correct for the time of transmission of the light-signal. He sets himself this problem—that man has travelled away from me at 161,000 miles a second for an unknown time x minutes; he has then sent a signal which travels the same distance back at 186,000 miles a second; the total time is 112 minutes; problem, find x. Answer, $x = 60$ minutes. He therefore judges that our cigar lasted 60 minutes, or twice as long as his own. His cigar lasted 30 minutes by his watch (because the same retardation affects both watch and cigar); and that was in our opinion twice as long as ours, because his watch was going at half-speed.

Here is the full time-table.

Stationary watch	Stationary Observer	Aviator	Aviator's watch
0 min.	Lights cigar	Lights cigar	0 min.
30 "	Finishes cigar	...	15 "
60 "	Inferred time aviator's cigar finished	Finishes cigar	30 "
112 "	Receives signal aviator's cigar finished	...	56 "
120 "	...	Inferred time stationary cigar finished	60 "
224 "	...	Receives signal stationary cigar finished	112 "

This is analysed from our point of view, not the aviator's; because it makes out that he was wrong in his inference and we were right. But no one can tell which was really right.

The argument will repay a careful examination, and it will be recognised that the chief cause of the paradox is that we assume that we are at rest in the aether, whereas the aviator assumes that he is at rest. Consequently whereas in our opinion the light-signal is overtaking him at merely the difference between 186,000 and 161,000 miles a second, he considers that it is coming to him through the relatively stationary aether at the normal speed of light. It must be remembered that each observer is furnished with complete experimental evidence in support of his own assumption. If we suggest to the aviator that owing to his high velocity the relative speed of the wave overtaking him can only be 25,000 miles a second, he will reply "I have determined the velocity of the wave relatively to me by timing it as it passes two points in my conveyance; and it turns out to be 186,000 miles a second. So I know my correction for light-time is right[2]." His clocks and scales are all behaving in an extraordinary way from our point of view, so it is not surprising that he should arrive at a measure of the velocity of the overtaking wave which differs from ours; but there is no way of convincing him that our reckoning is preferable.

Although not a very practical problem, it is of interest to inquire what happens when the aviator's speed is still further increased and approximates to the velocity of light. Lengths in the direction of flight become smaller and smaller, until for the speed of light they shrink to zero. The aviator and the objects accompanying him shrink to two dimensions. We are saved the difficulty of imagining how the processes of life can go on in two dimensions, because nothing goes on. Time is arrested altogether. This is the description according to the terrestrial observer. The aviator himself detects nothing unusual; he does not perceive that he has stopped moving. He is merely waiting for the next instant to come before making the next movement; and the mere fact that time is arrested means that he does not perceive that the next instant is a long time coming.

It is a favourite device for bringing home the vast distances of the stars to imagine a voyage through space with the velocity of light. The youthful adventurer steps on to his magic carpet loaded with provisions for a century. He reaches his journey's end, say Arcturus, a decrepit centenarian. This is wrong. It is quite true that the journey would last something like a hundred years by terrestrial chronology; but the adventurer would arrive at his destination no more aged than when he started, and he would not have had time to think of eating. So long as he travels with the speed of light he has immortality and eternal youth. If in some way his motion were reversed so that he returned to the earth again, he would find that centuries had elapsed here, whilst he himself did not feel a day older—for him the voyage had lasted only an instant[3].

---

[2]We need not stop to prove this directly. If the aviator could detect anything in his measurements inconsistent with the hypothesis that he was at rest in the aether (e.g. a difference of velocity of overtaking waves of light and waves meeting him) it would contradict the restricted principle of relativity.

[3]Since the earth is moving relatively to our adventurer with the velocity of light, we might be tempted to argue that from this point of view the terrestrial observer would have perpetual youth whilst the voyager grew older. Evidently, if they met again, they could disprove one or other of the two arguments. But in order to meet again the velocity of one of them must be

Our reason for discussing at length the effects of these improbably high velocities is simply in order that we may speak of the results in terms of common experience; otherwise it would be necessary to use the terms of refined technical measurement. The relativist is sometimes suspected of an inordinate fondness for paradox; but that is rather a misunderstanding of his argument. The paradoxes exist when the new experimental discoveries are woven into the scheme of physics hitherto current, and the relativist is ready enough to point this out. But the conclusion he draws is that a revised scheme of physics is needed in which the new experimental results will find a natural place without paradox.

To sum up—on any planet moving with a great velocity through the aether, extraordinary changes of length of objects are continually occurring as they move about, and there is a slowing down of all natural processes as though time were retarded. These things cannot be perceived by anyone on the planet; but similar effects would be detected by any observer having a great velocity relative to the planet (who makes all allowances for the effect of the motion on the observations, but takes it for granted that he himself is at rest in the aether[4]). There is complete reciprocity so that each of two observers in relative motion will find the same strange phenomena occurring to the other; and there is nothing to help us to decide which is right.

I think that no one can contemplate these results without feeling that the whole strangeness must arise from something perverse and inappropriate in our ordinary point of view. Changes go on on a planet, all nicely balanced by adjustments of natural forces, in such a way that no one on the planet can possibly detect what is taking place. Can we seriously imagine that there is anything in the reality behind the phenomena, which reflects these changes? Is it not more probable that we ourselves introduce the complexity, because our method of description is not well-adapted to give a simple and natural statement of what is really occurring?

The search for a more appropriate apparatus of description leads us to the standpoint of relativity described in the next chapter. I draw a distinction between the principle and the standpoint of relativity. The principle of relativity is a statement of experimental fact, which may be right or wrong; the first part of it—the restricted principle—has already been enunciated. Its consequences can be deduced by mathematical reasoning, as in the case of any other scientific generalization. It postulates no particular mechanism of nature, *and no particular view as to the meaning of time and space*, though it may suggest theories on the subject. The only question is whether it is experimentally true or not.

The standpoint of relativity is of a different character. It asserts first that certain unproved hypotheses as to time and space have insensibly crept into current physical theories, and that these are the source of the difficulties described

---

reversed by supernatural means or by an intense gravitational force so that the conditions are not symmetrical and reciprocity does not apply. The argument given in the text appears to be the correct one.

[4]The last clause is perhaps unnecessary. The correction applied for light transmission will naturally be based on the observer's own experimental determination of the velocity of light. According to experiment the velocity of light relatively to him is *apparently* the same in all directions, and he will apply the corrections accordingly. This is equivalent to assuming that he is at rest in the aether; but he need not, and probably would not, make the assumption explicitly.

above. Now the most dangerous hypotheses are those which are tacit and unconscious. So the standpoint of relativity proposes tentatively to do without these hypotheses (not making any others in their place); and it discovers that they are quite unnecessary and are not supported by any known fact. This in itself appears to be sufficient justification for the standpoint. Even if at some future time facts should be discovered which confirm the rejected hypotheses, the relativist is not wrong in reserving them until they are required.

It is not our policy to take shelter in impregnable positions; and we shall not hesitate to draw reasonable conclusions as well as absolutely proved conclusions from the knowledge available. But to those who think that the relativity theory is a passing phase of scientific thought, which may be reversed in the light of future experimental discoveries, we would point out that, though like other theories it may be developed and corrected, there is a certain minimum statement possible which represents irreversible progress. Certain hypotheses enter into all physical descriptions and theories hitherto current, dating back in some cases for 2000 years, in other cases for 200 years. It can now be proved that these hypotheses have nothing to do with any phenomena yet observed, and do not afford explanations of any known fact. This is surely a discovery of the greatest importance—quite apart from any question as to whether the hypotheses are actually wrong.

I am not satisfied with the view so often expressed that the sole aim of scientific theory is "economy of thought." I cannot reject the hope that theory is by slow stages leading us nearer to the truth of things. But unless science is to degenerate into idle guessing, the test of value of any theory must be whether it expresses with as little redundancy as possible the facts which it is intended to cover. Accidental truth of a conclusion is no compensation for erroneous deduction.

The relativity standpoint is then a discarding of certain hypotheses, which are uncalled for by any known facts, and stand in the way of an understanding of the simplicity of nature.

2 Relativity

The views of time and space, which I have to set forth, have their foundation in experimental physics. Therein is their strength. Their tendency is revolutionary. From henceforth space in itself and time in itself sink to mere shadows, and only a kind of union of the two preserves an independent existence.

H. Minkowski (1908).

THERE are two parties to every observation—the observed and the observer. What we see depends not only on the object looked at, but on our own circumstances—position, motion, or more personal idiosyncracies. Sometimes by instinctive habit, sometimes by design, we attempt to eliminate our own share in the observation, and so form a general picture of the world outside us, which shall be common to all observers. A small speck on the horizon of the sea is interpreted as a giant steamer. From the window of our railway carriage we see a cow glide past at fifty miles an hour, and remark that the creature is enjoying a rest. We see the starry heavens revolve round the earth, but decide that it is really the earth that is revolving, and so picture the state of the universe in a way which would be acceptable to an astronomer on any other planet.

The first step in throwing our knowledge into a common stock must be the elimination of the various individual standpoints and the reduction to some specified standard observer. The picture of the world so obtained is none the less relative. We have not eliminated the observer's share; we have only fixed it definitely.

To obtain a conception of the world from the point of view of no one in particular is a much more difficult task. The position of the observer can be eliminated; we are able to grasp the conception of a chair as an object in nature—looked at all round, and not from any particular angle or distance. We can think of it without mentally assigning ourselves some position with respect to it. This is a remarkable faculty, which has evidently been greatly assisted by the perception of solid relief with our two eyes. But the motion of the observer is not eliminated so simply. We had thought that it was accomplished; but the discovery in the last chapter that observers with different motions use different space- and time-reckoning shows that the matter is more complicated than was supposed. It may well require a complete change in our apparatus of description, because all the familiar terms of physics refer primarily to the relations of the world to an observer in some specified circumstances.

Whether we are able to go still further and obtain a knowledge of the world, which not merely does not particularise the observer, but does not postulate an observer at all; whether if such knowledge could be obtained, it would convey any intelligible meaning; and whether it could be of any conceivable interest to anybody if it could be understood—these questions need not detain us now. The answers are not necessarily negative, but they lie outside the normal scope of physics.

The circumstances of an observer which affect his observations are his position, motion and gauge of magnitude. More personal idiosyncracies disappear if, instead of relying on his crude senses, he employs scientific measuring apparatus. But scientific apparatus has position, motion and size, so that these are still involved in the results of any observation. There is no essential distinction between scientific measures and the measures of the senses. In either case our acquaintance with the external world comes to us through material channels; the observer's body can be regarded as part of his laboratory equipment, and, so far as we know, it obeys the same laws. We therefore group together perceptions and scientific measures, and in speaking of "a particular observer" we include all his measuring appliances.

Position, motion, magnitude-scale—these factors have a profound influence on the aspect of the world to us. Can we form a picture of the world which shall be a synthesis of what is seen by observers in all sorts of positions, having all sorts of velocities, and all sorts of sizes? As already stated we have accomplished the synthesis of positions. We have two eyes, which have dinned into our minds from babyhood that the world has to be looked at from more than one position. Our brains have so far responded as to give us the idea of solid relief, which enables us to appreciate the three-dimensional world in a vivid way that would be scarcely possible if we were only acquainted with strictly two-dimensional pictures. We not merely deduce the three-dimensional world; we see it. But we have no such aid in synthesising different motions. Perhaps if we had been endowed with two eyes moving with different velocities our brains would have developed the necessary faculty; we should have perceived a kind of relief in a fourth dimension so as to combine into one picture the aspect of things seen with different motions. Finally, if we had had two eyes of different sizes, we might have evolved a faculty for combining the points of view of the mammoth and the microbe.

It will be seen that we are not fully equipped by our senses for forming an impersonal picture of the world. And it is because the deficiency is manifest that we do not hesitate to advocate a conception of the world which transcends the images familiar to the senses. Such a world can perhaps be grasped, but not pictured by the brain. It would be unreasonable to limit our thought of nature to what can be comprised in sense-pictures. As Lodge has said, our senses were developed by the struggle for existence, not for the purpose of philosophising on the world.

Let us compare two well-known books, which might be described as elementary treatises on relativity, *Alice in Wonderland* and *Gulliver's Travels*. Alice was continually changing size, sometimes growing, sometimes on the point of vanishing altogether. Gulliver remained the same size, but on one occasion he encountered a race of men of minute size with everything in proportion, and on another voyage a land where everything was gigantic. It does not require much reflection to see that both authors are describing the same phenomenon—a relative change of scale of observer and observed. Lewis Carroll took what is probably the ordinary scientific view, that the observer had changed, rather than that a simultaneous change had occurred to all her surroundings. But it would never have appeared like that to Alice; she could not have "stepped outside and looked at herself," picturing herself as a giant filling the room. She would have said that the room had unaccountably shrunk. Dean Swift took the truer view of the human mind when he made Gulliver attribute his own changes to the things around him; it never occurred to Gulliver that his own size had altered; and, if he had thought of the explanation, he could scarcely have accustomed himself to that way of thinking. But both points of view are legitimate. The size of a thing can only be imagined as relative to something else; and there is no means of assigning the change to one end of the relation rather than the other.

We have seen in the theory of the Michelson-Morley experiment that, according to current physical views, our standard of size—the rigid measuring-rod—must change according to the circumstances of its motion; and the aviator's adventures illustrated a similar change in the standard of duration of time. Certain rather puzzling irregularities have been discovered in the apparent motions of the Sun, Mercury, Venus and the Moon; but there is a strong family resemblance between these, which leads us to believe that the real phenomenon is a failure of the time-keeping of our standard clock, the Earth. Instances could be multiplied where a change of the observer or his standards produces or conceals changes in the world around him.

The object of the relativity theory, however, is not to attempt the hopeless task of apportioning responsibility between the observer and the external world, but to emphasise that in our ordinary description and in our scientific description of natural phenomena the two factors are indissolubly united. All the familiar terms of physics—length, duration of time, motion, force, mass, energy, and so on—refer primarily to this relative knowledge of the world; and it remains to be seen whether any of them can be retained in a description of the world which is not relative to a particular observer.

Our first task is a description of the world independent of the motion of the observer. The question of the elimination of his gauge of magnitude belongs to a later development of the theory discussed in Chapter 11. Let us draw a square $ABCD$ on a sheet of paper, making the sides equal, to the best of our knowledge. We have seen that an aviator flying at $161,000$ miles a second in the direction AB, would judge that the sides AB, DC had contracted to half their length, so that for him the figure would be an oblong. If it were turned through a right angle AB and DC would expand and the other two sides contract—in his judgment. For us, the lengths of AB and AC are equal; for him, one length is twice the other. Clearly length cannot be a property inherent in our drawing; it needs the specification of some observer.

We have seen further that duration of time also requires that an observer should be specified. The stationary observer and the aviator disagreed as to whose cigar lasted the longer time.

Thus *length* and *duration* are not things inherent in the external world; they are relations of things in the external world to some specified observer. If we grasp this all the mystery disappears from the phenomena described in Chapter 1. When the rod in the Michelson-Morley experiment is turned through a right angle it contracts; that naturally gives the impression that something has happened to the rod itself. Nothing whatever has happened to the rod—the object in the external world. Its length has altered, but length is not an intrinsic property of the rod, since it is quite indeterminate until some observer is specified. Turning the rod through a right angle has altered the relation to the observer (implied in the discussion of the experiment); but the rod itself, or the relation of a molecule at one end to a molecule at the other, is unchanged. Measurement of length and duration is a comparison with partitions of space and time drawn by the observer concerned, with the help of apparatus which shares his motion. Nature is not concerned with these partitions; it has, as we shall see later, a geometry of its own which is of a different type.

Current physics has hitherto assumed that all observers are not to be regarded as on the same footing, and that there is some absolute observer whose judgments of length and duration are to be treated with respect, because nature pays attention to *his* space-time partitions. He is supposed to be at rest in the aether, and the aether materialises his space-partitions so that they have a real significance in the external world. This is sheer hypothesis, and we shall find it is unsupported by any facts. Evidently our proper course is to pursue our investigations, and call in this hypothetical observer only if we find there is something which he can help to explain.

We have been leading up from the older physics to the new outlook of relativity, and the reader may feel some doubt as to whether the strange phenomena of contraction and time-retardation, that were described in the last chapter, are to be taken seriously, or are part of a *reductio ad absurdum* argument. The answer is that we believe that the phenomena do occur as described; only the description (like that of all observed phenomena) concerns the relations of the external world to some observer, and not the external world itself. The startling character of the phenomena arises from the natural but fallacious inference that they involve intrinsic changes in the objects themselves.

We have been considering chiefly the observer's end of the observation; we must now turn to the other end—the thing observed. Although length and duration have no exact counterparts in the external world, it is clear that there is a certain ordering of things and events outside us which we must now find more appropriate terms to describe. The order of events is a four-fold order; we can arrange them as right-and-left, backwards-and-forwards, up-and-down, sooner-and-later. An individual may at first consider these as four independent orders, but he will soon attempt to combine some of them. It is recognised at once that there is no essential distinction between right-and-left and backwards-and-forwards. The observer has merely to turn through a right angle and the two are interchanged. If he turns through a smaller angle, he has first to combine them, and then to redivide them in a different way. Clearly it would be a nuisance to continually combine and redivide; so we get accustomed to the thought of leaving them combined in a two-fold or two-dimensional order. The amalgamation of up-and-down is less simple. There are obvious reasons for considering this dimension of the world as fundamentally distinct from the other two. Yet it would have been a great stumbling-block to science if the mind had refused to combine space into a three-dimensional whole. The combination has not concealed the real distinction of horizontal and vertical, but has enabled us to understand more clearly its nature—for what phenomena it is relevant, and for what irrelevant. We can understand how an observer in another country re-divides the combination into a different vertical and horizontal. We must now go further and amalgamate the fourth order, sooner-and-later. This is still harder for the mind. It does not imply that there is no distinction between space and time; but it gives a fresh unbiassed start by which to determine what the nature of the distinction is.

The idea of putting together space and time, so that time is regarded as a fourth dimension, is not new. But until recently it was regarded as merely a picturesque way of looking at things without any deep significance. We can put together time and temperature in a thermometer chart, or pressure and volume on an indicator-diagram. It is quite non-committal. But our theory is going to lead much further than that. We can lay two dimensional surfaces—sheets of paper—on one another till we build up a three-dimensional block; but there is a difference between a block which is a pile of sheets and a solid block of paper. The solid block is the true analogy for the four-dimensional combination of space-time; it does not separate naturally into a particular set of three-dimensional spaces piled in time-order. It can be redivided into such a pile; *but it can be redivided in any direction we please.*

Just as the observer by changing his orientation makes a new division of the two-dimensional plane into right-and-left, backwards-and-forwards—just as the observer by changing his longitude makes a new division of three-dimensional space into vertical and horizontal—so the observer by *changing his motion* makes a new division of the four-dimensional order into time and space.

This will be justified in detail later; it indicates that observers with different motions will have different time and space-reckoning—a conclusion we have already reached from another point of view.

Although different observers separate the four orders differently, they all agree that the order of events is four-fold; and it appears that this undivided four-fold order is the same for all observers. We therefore believe that it is inherent in the external world; it is in fact the synthesis, which we have been seeking, of the appearances seen by observers having all sorts of positions and all sorts of (uniform) motions. It is therefore to be regarded as a conception of the real world not relative to any particularly circumstanced observer.

The term "real world" is used in the ordinary sense of physics, without any intention of prejudging philosophical questions as to reality. It has the same degree of reality as was formerly attributed to the three-dimensional world of scientific theory or everyday conception, which by the advance of knowledge it replaces. As I have already indicated, it is merely the accident that we are not furnished with a pair of eyes in rapid relative motion, which has allowed our brains to neglect to develop a faculty for visualising this four-dimensional world as directly as we visualise its three-dimensional section.

It is now easy to see that length and duration must be the components of a single entity in the four-dimensional world of space-time. Just as we resolve a structure into plan and elevation, so we resolve extension in the four-dimensional world into length and duration. The structure has a size and shape independent of our choice of vertical. Similarly with things in space-time. Whereas length and duration are relative, the single "extension" of which they are components has an absolute significance in nature, independent of the particular decomposition into space and time separately adopted by the observer.

Consider two events; for example, the stroke of one o'clock and the stroke of two o'clock by Big Ben. These occupy two points in space-time, and there is a definite separation between them. An observer at Westminster considers that they occur at the same place, and that they are separated by an hour in time; thus he resolves their four-dimensional separation into zero distance in space and one hour distance in time. An observer on the sun considers that they do not occur at the same place; they are separated by about 70,000 miles, that being the distance travelled by the earth in its orbital motion with respect to the sun. It is clear that he is not resolving in quite the same directions as the terrestrial observer, since he finds the space-component to be 70,000 miles instead of zero. But if he alters one component he must necessarily alter the other; so he will make the time-component differ slightly from an hour. By analogy with resolution into components in three-dimensions, we should expect him to make it less than an hour—having, as it were, borrowed from time to make space; but as a matter of fact he makes it longer. This is because space-time has a different geometry, which will be described later. Our present point is that there is but one separation of two events in four dimensions, which can be resolved in any number of ways into the components length and duration.

We see further how motion must be purely relative. Take two events A and B in the history of one particle. We can choose any direction as the time-direction; let us choose it along AB. Then A and B are separated only in time and not in space, so the particle is at rest. If we choose a slightly inclined time-direction, the separation AB will have a component in space; the two events then do not occur at the same place, that is to say, the particle has moved. The negation of absolute motion is thus associated with the possibility of choosing the time-direction in any way we please. What determines the separation of space and time for any particular observer can now be seen. Let the observer place himself so that he is, to the best of his knowledge, at rest. If he is a normal human being, he will seat himself in an arm-chair; if he is an astronomer, he will place himself on the sun or at the centre of the stellar universe. Then all the events happening directly to him will in his opinion occur at the same place. Their separation will have no space-component, and they will accordingly be ranged solely in the time-direction. This chain of events, marking his track through the four-dimensional world, will be his time-direction. Each observer bases his separation of space and time on his own track through the world.

Since any separation of space and time is admissible, it is possible for the astronomer to base his space and time on the track of a solar observer instead of that of a terrestrial observer; but it must be remembered that in practice the space and time of the solar observer have to be inferred indirectly from those of the terrestrial observer; and, if the corrections are made according to the crude methods hitherto employed, they may be inferred wrongly (if extreme accuracy is needed).

The most formidable objection to this relativist view of the world is the aether difficulty. We have seen that uniform motion through the aether cannot be detected by experiment, and therefore it is entirely in accordance with experiment that such motion should have no counterpart in the four-dimensional world. Nevertheless, it would almost seem that such motion must logically exist, if the aether exists; and, even at the expense of formal simplicity, it ought to be exhibited in any theory which pretends to give a complete account of what is going on in nature. If a substantial aether analogous to a material ocean exists, it must rigidify, as it were, a definite space; and whether the observer or whether nature pays any attention to that space or not, a fundamental separation of space and time must be there. Some would cut the knot by denying the aether altogether. We do not consider that desirable, or, so far as we can see, possible; but we do deny that the aether need have such properties as to separate space and time in the way supposed. It seems an abuse of language to speak of a division existing, when nothing has ever been found to pay any attention to the division.

Mathematicians of the nineteenth century devoted much time to theories of elastic solid and other material aethers. Waves of light were supposed to be actual oscillations of this substance; it was thought to have the familiar properties of rigidity and density; it was sometimes even assigned a place in the table of the elements. The real death-blow to this materialistic conception of the aether was given when attempts were made to explain matter as some state in the aether. For if matter is vortex-motion or beknottedness in aether, the aether cannot be matter–some state in itself. If any property of matter comes to be regarded as a thing to be explained by a theory of its structure, clearly that property need not be attributed to the aether. If physics evolves a theory of matter which explains some property, it stultifies itself when it postulates that the same property exists unexplained in the primitive basis of matter.

Moreover the aether has ceased to take any very active part in physical theory and has, as it were, gone into reserve. A modern writer on electromagnetic theory will generally start with the postulate of an aether pervading all space; he will then explain that at any point in it there is an electromagnetic vector whose intensity can be measured; henceforth his sole dealings are with this vector, and probably nothing more will be heard of the aether itself. In a vague way it is supposed that this vector represents some condition of the aether, and we need not dispute that without some such background the vector would scarcely be intelligible—but the aether is now only a background and not an active participant in the theory.

There is accordingly no reason to transfer to this vague background of aether the properties of a material ocean. Its properties must be determined by experiment, not by analogy. In particular there is no reason to suppose that it can partition out space in a definite way, as a material ocean would do. We have seen in the Prologue that natural geometry depends on laws of matter; therefore it need not apply to the aether. Permanent identity of particles is a property of matter, which Lord Kelvin sought to explain in his vortex-ring hypothesis. This abandoned hypothesis at least teaches us that permanence should not be regarded as axiomatic, but may be the result of elaborate constitution. There need not be anything corresponding to permanent identity in the constituent portions of the aether; we cannot lay our finger at one spot and say "this piece of aether was a few seconds ago over there." Without any continuity of identity of the aether motion through the aether becomes meaningless; and it seems likely that this is the true reason why no experiment ever reveals it.

This modern theory of the relativity of all uniform motion is essentially a return to the original Newtonian view, temporarily disturbed by the introduction of aether problems; for in Newton's dynamics uniform motion of the whole system has not—and no one would expect it to have—any effect. But there are considerable difficulties in the limitation to uniform motion. Newton himself seems to have appreciated the difficulty; but the experimental evidence appeared to him to be against any extension of the principle. Accordingly Newton's laws of mechanics are not of the general type in which it is unnecessary to particularise the observer; they hold only for observers with a special kind of motion which is described as "unaccelerated." The only definition of this epithet that can be given is that an "unaccelerated" observer is one for whom Newton's laws of motion hold. On this theory, the phenomena are not indifferent to an acceleration or non-uniform motion of the whole system. Yet an absolute non-uniform motion through space is just as impossible to imagine as an absolute uniform motion. The partial relativity of phenomena makes the difficulty all the greater. If we deny a fundamental medium with continuous identity of its parts, motion uniform or non-uniform should have no significance; if we admit such a medium, motion uniform or non-uniform should be detectable; but it is much more difficult to devise a plan of the world according to which uniform motion has no significance and non-uniform motion is significant.

It is through experiment that we have been led back to the principle of relativity for uniform motion. In seeking some kind of extension of this principle to accelerated motion, we are led by the feeling that, having got so far, it is difficult and arbitrary to stop at this point. We now try to conceive a system of nature for which all kinds of motion of the observer are indifferent. It will be a completion of our synthesis of what is perceived by observers having all kinds of motions with respect to one another, removing the restriction to uniform motion. The experimental tests must follow after the consequences of this generalisation have been deduced.

The task of formulating such a theory long appeared impossible. It was pointed out by Newton that, whereas there is no criterion for detecting whether a body is at rest or in uniform motion, it is easy to detect whether it is in rotation. For example the bulge of the earth's equator is a sign that the earth is rotating, since a plastic body at rest would be spherical.

This problem of rotation affords a hint as to the cause of the incomplete relativity of Newtonian mechanics. The laws of motion are formulated with respect to an unaccelerated observer, and do not apply to a frame of reference rotating with the earth. Yet mathematicians frequently do use a rotating frame. Some modification of the laws is then necessary; and the modification is made by introducing a centrifugal force—not regarded as a real force like gravitation, but as a mathematical fiction employed to correct for the improper choice of a frame of reference. The bulge of the earth's equator may be attributed indifferently to the earth's rotation or to the outward pull of the centrifugal force introduced when the earth is regarded as non-rotating.

Now it is generally assumed that the centrifugal force is something *sui generis*, which could always be distinguished experimentally from any other natural phenomenon. If then on choosing a frame of reference we find that a centrifugal force is detected, we can at once infer that the frame of reference is a "wrong" one; rotating and non-rotating frames can be distinguished by experiment, and rotation is thus strictly absolute. But this assumes that the observed effects of centrifugal force cannot be produced in any other way than by rotation of the observer's frame of reference. If once it is admitted that centrifugal force may not be completely distinguishable by experiment from another kind of force—gravitation—perceived even by Newton's unaccelerated observer, the argument ceases to apply. We can never determine exactly how much of the observed field of force is centrifugal force and how much is gravitation; and we cannot find experimentally any definite standard that is to be considered absolutely non-rotating.

The question then, whether there exists a distinction between "right" frames of reference and "wrong" frames, turns on whether the use of a "wrong" frame produces effects experimentally distinguishable from any natural effects which can be perceived when a "right" frame is used. If there is no such difference, all frames may be regarded as on the same footing and equally right. In that case we can have a complete relativity of natural phenomena. Since the effect of departing from Newton's standard frame is the introduction of a field of force, this generalised relativity theory must be largely occupied with the nature of fields of force.

The precise meaning of the statement that all frames of reference are on the same footing is rather difficult to grasp. We believe that there are absolute things in the world—not only matter, but certain characteristics in empty space or aether. In the atmosphere a frame of reference which moves with the air is differentiated from other frames moving in a different manner; this is because, besides discharging the normal functions of a frame of reference, the air-frame embodies certain of the absolute properties of the matter existing in the region. Similarly, if in empty space we choose a frame of reference which more or less follows the lines of the absolute structure in the region, the frame will usurp some of the absolute qualities of that structure. What we mean by the equivalence of all frames is that they are not differentiated by any qualities formerly supposed to be intrinsic in the frames themselves—rest, rectangularity, acceleration—independent of the absolute structure of the world that is referred to them. Accordingly the objection to attributing absolute properties to Newton's frame of reference is not that it is impossible for a frame of reference to acquire absolute properties, but that the Newtonian frame has been laid down on the basis of relative knowledge without any attempt to follow the lines of absolute structure.

Force, as known to us observationally, is like the other quantities of physics, a relation. The force, measured with a spring-balance, for example, depends on the acceleration of the observer holding the balance; and the term may, like length and duration, have no exact counterpart in a description of nature independent of the observer. Newton's view assumes that there is such a counterpart, an active cause in nature which is identical with the force perceived by his standard unaccelerated observer. Although any other observer perceives this force with additions of his own, it is implied that the original force in nature and the observer's additions can in some way be separated without ambiguity. There is no experimental foundation for this separation, and the relativity view is that a field of force can, like length and duration, be nothing but a link between nature and the observer. There is, of course, something at the far end of the link, just as we found an extension in four dimensions at the far end of the relations denoted by length and duration. We shall have to study the nature of this unknown whose relation to us appears as force. Meanwhile we shall realise that the alteration of perception of force by non-uniform motion of the observer, as well as the alteration of the perception of length by his uniform motion, is what might be expected from the nature of these quantities as relations solely.

We proceed now to a more detailed study of the four-dimensional world, of the things which occur in it, and of the laws by which they are regulated. It is necessary to dive into this absolute world to seek the truth about nature; but the physicist's object is always to obtain knowledge which can be applied to the relative and familiar aspect of the world. The absolute world is of so different a nature, that the relative world, with which we are acquainted, seems almost like a dream. But if indeed we are dreaming, our concern is with the baseless fabric of our vision. We do not suggest that physicists ought to translate their results into terms of four-dimensional space for the empty satisfaction of working in the realm of reality. It is rather the opposite. They explore the new field and bring back their spoils—a few simple generalisations—to apply them to the practical world of three-dimensions. Some guiding light will be given to the attempts to build a scheme of things entire. For the rest, physics will continue undisturbedly to explore the relative world, and to employ the terms applicable to relative knowledge, but with a fuller appreciation of its relativity.

3 The World of Four Dimensions

Here is a portrait of a man at eight years old, another at fifteen, another at seventeen, another at twenty-three, and so on. All these are evidently sections, as it were, Three-Dimensional representations of his Four-Dimensional being, which is a fixed and unalterable thing.

H. G. Wells, *The Time Machine.*

THE distinction between horizontal and vertical is not an illusion; and the man who thinks it can be disregarded is likely to come to an untimely end. Yet we cannot arrive at a comprehensive view of nature unless we combine horizontal and vertical dimensions into a three-dimensional space. By doing this we obtain a better idea of what the distinction of horizontal and vertical really is in those cases where it is relevant, e.g. the phenomena of motion of a projectile. We recognise also that vertical is not a universally differentiated direction in space, as the flat-earth philosophers might have imagined.

Similarly by combining the time-ordering and space-ordering of the events of nature into a single order of four dimensions, we shall not only obtain greater simplicity for the phenomena in which the separation of time and space is irrelevant, but we shall understand better the nature of the differentiation when it is relevant.

A point in this space-time, that is to say a given instant at a given place, is called an "event." An event in its customary meaning would be the physical happening which occurs at and identifies a particular place and time. However, we shall use the word in both senses, because it is scarcely possible to think of a point in space-time without imagining some identifying occurrence.

In the ordinary geometry of two or three dimensions, the distance between two points is something which can be measured, usually with a rigid scale; it is supposed to be the same for all observers, and there is no need to specify horizontal and vertical directions or a particular system of coordinates. In four-dimensional space-time there is likewise a certain extension or generalised distance between two events, of which the distance in space and the separation in time are particular components. This extension in space and time combined is called the "interval" between the two events; it is the same for all observers, however they resolve it into space and time separately. We may think of the interval as something intrinsic in external nature—an absolute relation of the two events, which postulates no particular observer. Its practical measurement is suggested by analogy with the distance of two points in space.

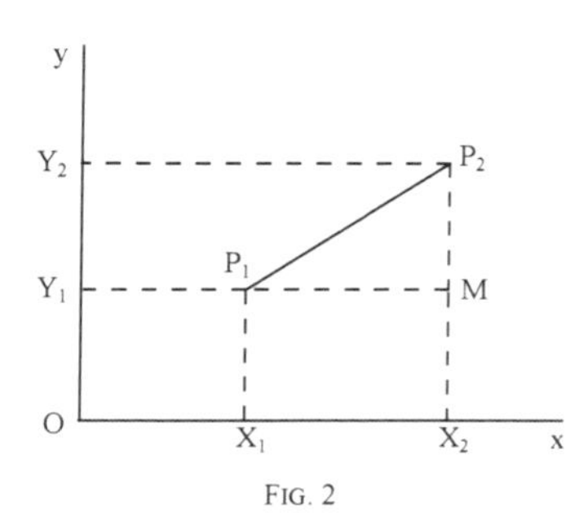

FIG. 2

In two dimensions on a plane, two points P_1, P_2 Fig. 2 can be specified by their rectangular coordinates (x_1, y_1) and (x_2, y_2), when arbitrary axes have been selected. In the figure, $OX_1 = x_1$, $OY_1 = y_1$, etc. We have

$$\begin{aligned} P_1P_2{}^2 &= P_1M^2 + MP_2{}^2 \\ &= X_1X_2{}^2 + Y_1Y_2{}^2 \\ &= (x_2 - x_1)^2 + (y_2 - y_1)^2, \end{aligned}$$

so that if s is the distance between P_1 and P_2

$$s^2 = (x_2 - x_1)^2 + (y_2 - y_1)^2.$$

The extension to three dimensions is, as we should expect,

$$s^2 = (x_2 - x_1)^2 + (y_2 - y_1)^2 + (z_2 - z_1)^2.$$

Introducing the times of the events t_1, t_2, we should naturally expect that the interval in the four-dimensional world would be given by

$$s^2 = (x_2 - x_1)^2 + (y_2 - y_1)^2 + (z_2 - z_1)^2 + (t_2 - t_1)^2.$$

An important point arises here. It was, of course, assumed that the same scale was used for measuring x and y and z. But how are we to use the same scale for measuring t? We cannot use a scale at all; some kind of clock is needed. The most natural connection between the measure of time and length is given by the fact that light travels 300,000 kilometres in 1 second. For the four-dimensional world we shall accordingly regard 1 second as the equivalent of 300,000 kilometres, and measure lengths and times in seconds or kilometres indiscriminately; in other words we make the velocity of light the unit of velocity. It is not essential to do this, but it greatly simplifies the discussion.

Secondly, the formulae here given for s^2 are the characteristic formulae of Euclidean geometry. So far as three-dimensional space is concerned the applicability of Euclidean geometry is very closely confirmed by experiment. But space-time is not Euclidean; it does, however, conform (at least approximately) to a very simple modification of Euclidean geometry indicated by the corrected formula

$$s^2 = (x_2 - x_1)^2 + (y_2 - y_1)^2 + (z_2 - z_1)^2 - (t_2 - t_1)^2.$$

There is only a sign altered; but that minus sign is the secret of the differences of the manifestations of time and space in nature.

This change of sign is often found puzzling at the start. We could not define s by the expression originally proposed (with the positive sign), because the expression does not define anything objective. Using the space and time of one observer, one value is obtained; for another observer, another value is obtained. But if s is defined by the expression now given, it is found that the same result is obtained by all observers[1]. The quantity s is thus something which concerns

[1] Appendix, Note 2.

solely the two events chosen; we give it a name—the interval between the two events. In ordinary space the distance between two points is the corresponding property, which concerns only the two points and not the extraneous coordinate system of location which is used. Hence interval, as here defined, is the analogue of distance; and the analogy is strengthened by the evident resemblance of the formula for s in both cases. Moreover, when the difference of time vanishes, the interval reduces to the distance. But the discrepancy of sign introduces certain important differences. These differences are summed up in the statement that the geometry of space is Euclidean, but the geometry of space-time is semi-Euclidean or "hyperbolic." The association of a geometry with any continuum always implies the existence of some uniquely measurable quantity like interval or distance; in ordinary space, geometry without the idea of distance would be meaningless.

For the moment the difficulty of thinking in terms of an unfamiliar geometry may be evaded by a dodge. Instead of real time t, consider imaginary time τ; that is to say, let

$$t = \tau\sqrt{-1}.$$

Then

$$(t_2 - t_1)^2 = -(\tau_2 - \tau_1)^2,$$

so that

$$s^2 = (x_2 - x_1)^2 + (y_2 - y_1)^2 + (z_2 - z_1)^2 + (\tau_2 - \tau_1)^2.$$

Everything is now symmetrical and there is no distinction between τ and the other variables. The continuum formed of space and imaginary time is completely isotropic for all measurements; no direction can be picked out in it as fundamentally distinct from any other.

The observer's separation of this continuum into space and time consists in slicing it in some direction, viz. that perpendicular to the path along which he is himself travelling. The section gives three-dimensional space at some moment, and the perpendicular dimension is (imaginary) time. Clearly the slice may be taken in any direction; there is no question of a true separation and a fictitious separation. There is no conspiracy of the forces of nature to conceal our absolute motion—because, looked at from this broader point of view, there is nothing to conceal. The observer is at liberty to orient his rectangular axes of x, y, z and τ arbitrarily, just as in three-dimensions he can orient his axes of x, y, z arbitrarily.

It can be shown that the different space and time used by the aviator in Chapter 1 correspond to an orientation of the time-axis along his own course in the four-dimensional world, whereas the ordinary time and space are given when the time-axis is oriented along the course of a terrestrial observer. The FitzGerald contraction and the change of time-measurement are given exactly by the usual formulae for rotation of rectangular axes[2].

It is not very profitable to speculate on the implication of the mysterious factor $\sqrt{-1}$, which seems to have the property of turning time into space. It can scarcely be regarded as more than an analytical device. To follow out the theory of the four-dimensional world in more detail, it is necessary to return to real time, and face the difficulties of a strange geometry.

[2]Appendix, Note 3.

Consider a particular observer, S, and represent time according to his reckoning by distance up the page parallel to OT. One dimension of his space will be represented by horizontal distance parallel to OX; another will stand out at right angles from the page; and the reader must imagine the third as best he can. Fortunately it will be sufficient for us to consider only the one dimension of space OX and deal with the phenomena of "line-land," i.e. we limit ourselves to motion to and fro in one straight line in space.

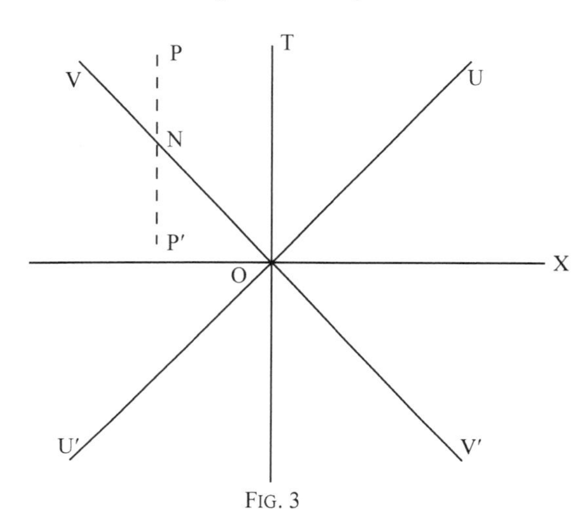

FIG. 3

The two lines $U'OU$, $V'OV$, at 45 to the axes, represent the tracks of points which progress 1 unit horizontally (in space) for 1 unit vertically (in time); thus they represent points moving with unit velocity. We have chosen the velocity of light as unit velocity; hence $U'OU$, $V'OV$ will be the tracks of pulses of light in opposite directions along the straight line.

Any event P within the sector UOV is indubitably after the event O, whatever system of time-reckoning is adopted. For it would be possible for a material particle to travel from O to P, the necessary velocity being less than that of light; and no rational observer would venture to state that the particle had completed its journey before it had begun it. It would, in fact, be possible for an observer travelling along NP to receive a light-signal or wireless telegram announcing the event O, just as he reached N, since ON is the track of such a message; and then after the time NP he would have direct experience of the event P. To have actual evidence of the occurrence of one event before experiencing the second is a clear proof of their absolute order in nature, which should convince not merely the observer concerned but any other observer with whom he can communicate.

Similarly events in the sector $U'OV'$ are indubitably before the event O.

With regard to an event P' in the sector UOV' or VOU' we cannot assert that it is absolutely before or after O. According to the time-reckoning of our chosen observer S, P' is after O, because it lies above the line OX; but there is nothing absolute about this. The track OP' corresponds to a velocity greater than that of light, so that we know of no particle or physical impulse which could follow the track. An observer experiencing the event P' could not get news of the event O by any known means until after P' had happened. The order of the two events can therefore only be inferred by estimating the delay of the message and this estimate will depend on the observer's mode of reckoning space and time.

Space-time is thus divided into three zones with respect to the event O. $U'OV'$ belongs to the indubitable past. UOV is the indubitable future. UOV' and VOU' are (absolutely) neither past nor future, but simply "elsewhere." It may be remarked that, as we have no means of identifying points in space as "the same point," and as the events O and P might quite well happen to the same particle of matter, the events are not necessarily to be regarded as in different places, though the observer S will judge them so; but the events O and P' cannot happen to the same particle, and no observer could regard them as happening at the same place. The main interest of this analysis is that it shows that the arbitrariness of time-direction is not inconsistent with the existence of regions of absolute past and future.

Although there is an absolute past and future, there is between them an extended neutral zone; and simultaneity of events at different places has no absolute meaning. For our selected observer all events along OX are simultaneous with one another; for another observer the line of events simultaneous with O would lie in a different direction. The denial of absolute simultaneity is a natural complement to the denial of absolute motion. The latter asserts that we cannot find out what is the same place at two different times; the former that we cannot find out what is the same time at two different places. It is curious that the philosophical denial of absolute motion is readily accepted, whilst the denial of absolute simultaneity appears to many people revolutionary.

The division into past and future (a feature of time-order which has no analogy in space-order) is closely associated with our ideas of causation and free will. In a perfectly determinate scheme the past and future may be regarded as lying mapped out—as much available to present exploration as the distant parts of space. Events do not happen; they are just there, and we come across them. "The formality of taking place" is merely the indication that the observer has on his voyage of exploration passed into the absolute future of the event in question; and it has no important significance. We can be aware of an eclipse in the year 1999, very much as we are aware of an unseen companion to Algol. Our knowledge of things *where* we are not, and of things *when* we are not, is essentially the same—an inference (sometimes a mistaken inference) from brain impressions, including memory, *here* and *now*.

So, if events are determinate, there is nothing to prevent a person from being *aware* of an event before it happens; and an event may cause other events previous to it. Thus the eclipse of the Sun in May 1919 caused observers to embark in March. It may be said that it was not the eclipse, but the calculations of the eclipse, which caused the embarkation; but I do not think any such distinction is possible, having regard to the indirect character of our acquaintance with all events except those at the precise point of space where we stand. A detached observer contemplating our world would see some events apparently causing events in their future, others apparently causing events in their past—the truth being that all are linked by determinate laws, the so-called causal events being merely conspicuous foci from which the links radiate.

The recognition of an absolute past and future seems to depend on the possibility of events which are not governed by a determinate scheme. If, say, the event O is an ultimatum, and the person describing the path NP is a ruler of the country affected, then it may be manifest to all observers that it is his knowledge of the actual occurrence of the event O which has caused him to create the event P. P must then be in the absolute future of O, and, as we have seen, must

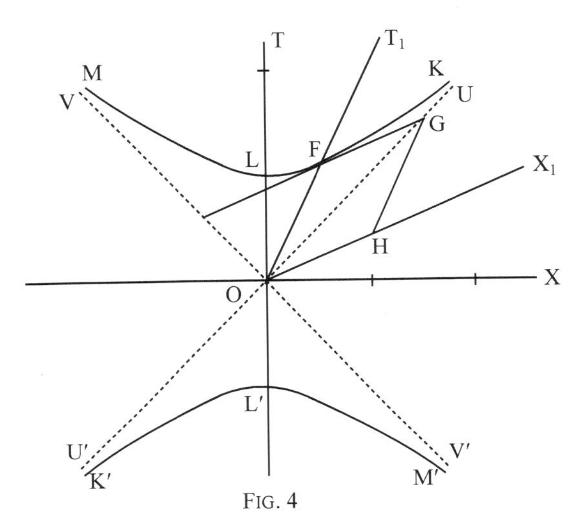

FIG. 4

lie in the sector UOV. But the inference is only permissible, if the event P could be determined by the event O, and was not predetermined by causes anterior to both—if it was possible for it to happen or not, consistently with the laws of nature. Since physics does not attempt to cover indeterminate events of this kind, the distinction of absolute past and future is not directly important for physics; but it is of interest to show that the theory of four-dimensional space-time provides an absolute past and future, in accordance with common requirements, although this can usually be ignored in applications to physics.

Consider now all the events which are at an interval of one unit from O, according to the definition of the interval s

$$s^2 = -(x_2 - x_1)^2 - (y_2 - y_1)^2 - (z_2 - z_1)^2 + (t_2 - t_1)^2. \tag{1}$$

We have changed the sign of s^2, because usually (though not always) the original s^2 would have come out negative. In Euclidean space points distant a unit interval lie on a circle; but, owing to the change in geometry due to the altered sign of $(t_2 - t_1)^2$, they now lie on a rectangular hyperbola with two branches KLM, $K'L'M'$. Since the interval is an absolute quantity, all observers will agree that these points are at unit interval from O.

Now make the following construction:—draw a straight line OFT_1 to meet the hyperbola in F; draw the tangent FG at F, meeting the light-line $U'OU$ in G; complete the parallelogram $OFGH$; produce OH to X_1. We now assert that an observer S_1 who chooses OT_1 for his time-direction will regard OX_1 as his space direction and will consider OF and OH to be the units of time and space.

The two observers make their partitions of space and time in different ways, as illustrated in Figs. 5 and 6, where in each case the partitions are at unit distance (in space and time) according to the observers' own reckoning. The same diagram of events in the world will serve for both observers; S_1 merely removes S's partitions and overlays his own, locating the events in his space and time accordingly. It will be seen at once that the lines of unit velocity—progress of one unit of space for one unit of time—agree, so that the velocity of a pulse of light is unity for both observers. It can be shown from the properties of the hyperbola that the locus of points at any interval s from O, given by equation (1), viz.

$$s^2 = (t - t_0)^2 - (x - x_0)^2,$$

is the same locus (a hyperbola) for both systems of reckoning x and t. The two observers will always agree on the measures of intervals, though they will disagree about lengths, durations, and the velocities of everything except light. This rather complex transformation is mathematically equivalent to the simple rotation of the axes required when imaginary time is used. It must not be

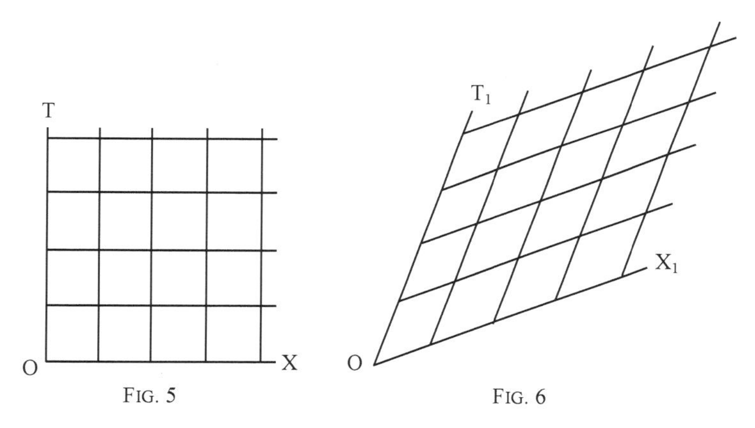

FIG. 5

FIG. 6

supposed that there is any natural distinction corresponding to the difference between the square-partitions of observer S and the diamond-shaped partitions of observer S_1. We might say that S_1 transplants the space-time world unchanged from Fig. 5 to Fig. 6, and then distorts it until the diamonds shown become squares; or we might equally well start with this distorted space-time, partitioned by S_1 into squares, and then S's partitions would be represented by diamonds. It cannot be said that either observer's space-time is distorted absolutely, but one is distorted relatively to the other. It is the relation of *order* which is intrinsic in nature, and is the same both for the squares and diamonds; *shape* is put into nature by the observer when he has chosen his partitions.

We can now deduce the FitzGerald contraction. Consider a rod of unit length at rest relatively to the observer S. The two extremities are at rest in his space, and consequently remain on the same space-partitions; hence their tracks in four dimensions PP', QQ' (Fig. 7) are entirely in the time-direction. The real rod in nature is the four-dimensional object shown in section as $P'PQQ'$. Overlay the same figure with S_1's space and time partitions, shown by the dotted lines. Taking a section at any one "time," the instantaneous rod is P_1Q_1, viz. the section of $P'PQQ'$ by S_1's time-line. Although on paper P_1Q_1 is actually longer than PQ, it is seen that it is a little shorter than one of S_1's space-partitions; and accordingly S_1 judges that it is less than one unit long—it has contracted on account of its motion relative to him.

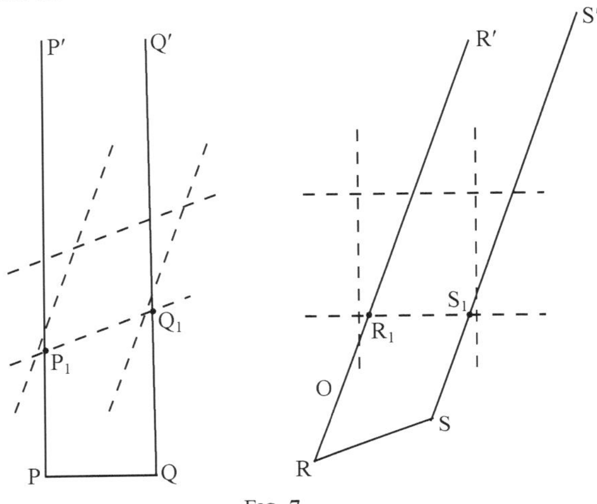

FIG. 7

Similarly $RR' - SS'$ is a rod of unit length at rest relatively to S_1. Overlaying S's partitions we see that it occupies R_1S_1 at a particular instant for S; and this is less than one of S's partitions. Thus S judges it to have contracted on account of its motion relative to him.

In the same way we can illustrate the problem of the duration of the cigar; each observer believed the other's cigar to last the longer time. Taking LM (Fig. 8) to represent the duration of S's cigar (two units), we see that in S_1's reckoning it reaches over a little more than two time-partitions. Moreover it has not kept to one space-partition, i.e. it has moved. Similarly $L'N'$ is the duration of S_1's cigar (two time-units for him); and it lasts a little beyond two unit-partitions in S's time-reckoning. (Note, in comparing the two diagrams, L', M', N' are the same points as L, M, N.)

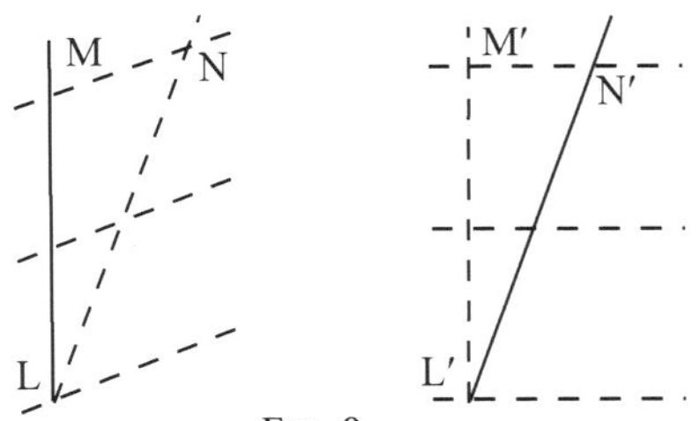

M
N
L
M′
N′
L′

FIG. 8

If in Fig. 4 we had taken the line OT_1 very near to OU, our diamonds would have been very elongated, and the unit-divisions OF, OH very large. This kind of partition would be made by an observer whose course through the world is OT_1, and who is accordingly travelling with a velocity approaching that of light relative to S. In the limit, when the velocity reaches that of light, both space-unit and time-unit become infinite, so that in the natural units for an observer travelling with the speed of light, all the events in the finite experience of S take place "in no time" and the size of every object is zero. This applies, however, only to the two dimensions x and t; the space-partitions parallel to the plane of the paper are not affected by this motion along x. Consequently for an observer travelling with the speed of light all ordinary objects become two-dimensional, preserving their lateral dimensions, but infinitely thin longitudinally. The fact that events take place "in no time" is usually explained by saying that the inertia of any particle moving with the velocity of light becomes infinite so that all molecular processes in the observer must stop; many things may happen in S's world in a twinkling of an eye—of S_1's eye.

However successful the theory of a four-dimensional world may be, it is difficult to ignore a voice inside us which whispers "At the back of your mind, you know that a fourth dimension is all nonsense." I fancy that that voice must often have had a busy time in the past history of physics. What nonsense to say that this solid table on which I am writing is a collection of electrons moving with prodigious speeds in empty spaces, which relatively to electronic dimensions are as wide as the spaces between the planets in the solar system! What nonsense to say that the thin air is trying to crush my body with a load of 14 lbs. to the square inch! What nonsense that the star-cluster, which I see through the telescope obviously there now, is a glimpse into a past age 50, 000 years ago! Let us not be beguiled by this voice. It is discredited.

But the statement that time is a fourth dimension may suggest unnecessary difficulties which a more precise definition avoids. It is in the external world that the four dimensions are united—not in the relations of the external world to the individual which constitute his direct acquaintance with space and time. Just in that process of relation to an individual, the order falls apart into the distinct manifestations of space and time. An individual is a four-dimensional object of greatly elongated form; in ordinary language we say that he has considerable extension in time and insignificant extension in space. Practically he is represented by a line—his track through the world. When the world is related to such an individual, his own asymmetry is introduced into the relation; and that order of events which is parallel with his track, that is to say with *himself*, appears in his experience to be differentiated from all other orders of events.

Probably the best known exposition of the fourth dimension is that given in E. Abbott's popular book *Flatland*. It may be of interest to see how far the four-dimensional world of space-time conforms with his anticipations. He lays stress on three points.

(1) As a four-dimensional body moves, its section by the three-dimensional world may vary; thus a rigid body can alter size and shape.

(2) It should be possible for a body to enter a completely closed room, by travelling into it in the direction of the fourth dimension, just as we can bring our pencil down on to any point within a square without crossing its sides.

(3) It should be possible to see the inside of a solid, just as we can see the inside of a square by viewing it from a point outside its plane.

The first phenomenon is manifested by the FitzGerald contraction.

If quantity of matter is to be identified with its mass, the second phenomenon does not happen. It could easily be conceived of as happening, but it is provided against by a special law of nature—the conservation of mass. It could happen, but it does not happen.

The third phenomenon does not happen for two reasons. A natural body extends in time as well as in space, and is therefore four-dimensional; but for the analogy to hold, the object must have one dimension less than the world, like the square seen from the third dimension. If the solid suddenly went out of existence so as to present a plane section towards time, we should still fail to see the interior of it; because light-tracks in four-dimensions are restricted to certain lines like $UOV, U'OV'$ in Fig. 3, whereas in three-dimensions light can traverse any straight line. This could be remedied by interposing some kind of dispersive medium, so that light of some wave-length could be found travelling with every velocity and following every track in space-time; then, looking at a solid which suddenly went out of existence, we should receive at the same moment light-impressions from every particle in its interior (supposing them self-luminous). We actually should see the inside of it.

How our poor eyes are to disentangle this overwhelming experience is quite another question.

The interval is a quantity so fundamental for us that we may consider its measurement in some detail. Suppose we have a scale AB divided into kilometres, say, and at each division is placed a clock also registering kilometres. (It will be remembered that time can be measured in seconds or kilometres indifferently.) When the clocks are correctly set and viewed from A the sum of the readings of

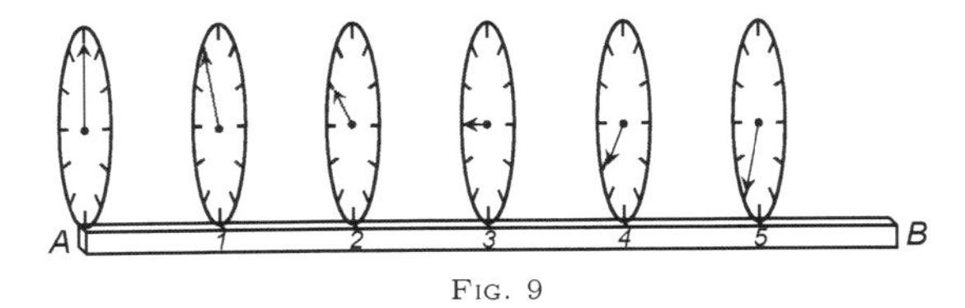

FIG. 9

any clock and the division beside it is the same for all, since the scale-reading gives the correction for the time taken by light, travelling with unit velocity, to reach A. This is shown in Fig. 9 where the clock-readings are given as though they were being viewed from A.

Now lay the scale in line with the two events; note the clock and scale-readings t_1, x_1, of the first event, and the corresponding readings t_2, x_2, of the second event. Then by the formula already given

$$s^2 = (t_2 - t_1)^2 - (x_2 - x_1)^2.$$

But suppose we took a different standard of rest, and set the scale moving uniformly in the direction AB. Then the divisions would have advanced to meet the second event, and $(x_2 - x_1)$ would be smaller. This is compensated, because $t_2 - t_1$ also becomes altered. A is now advancing to meet the light coming from any of the clocks along the rod; the light arrives too quickly, and in the initial adjustment described above the clock must be set back a little. The clock-reading of the event is thus smaller. There are other small corrections arising from the

FitzGerald contraction, etc.; and the net result is that, it does not matter what uniform motion is given to the scale, the final result for s is always the same.

In elementary mechanics we are taught that velocities can be compounded by adding. If B's velocity relative to A (as observed by either of them) is 100 km. per sec., and C's velocity relative to B is 100 km. per sec. in the same direction, then C's velocity relative to A should be 200 km. per sec. This is not quite accurate; the true answer is 199.999978 km. per sec. The discrepancy is not difficult to explain. The two velocities and their resultant are not all reckoned with respect to the same partitions of space and time. When B measures C's velocity relative to him he uses his own space and time, and it must be corrected to reduce to A's space and time units, before it can be added on to a velocity measured by A.

If we continue the chain, introducing D whose velocity relative to C, and measured by C, is 100 km. per sec., and so on *ad infinitum*, we never obtain an infinite velocity with respect to A, but gradually approach the limiting velocity of $300,000$ km. per sec., the speed of light. This speed has the remarkable property of being absolute, whereas every other speed is relative. If a speed of 100 km. per sec. or of $100,000$ km. per sec. is mentioned, we have to ask—speed relative to what? But if a speed of $300,000$ km. per sec. is mentioned, there is no need to ask the question; the answer is—relative to any and every piece of matter. A β particle shot off from radium can move at more than $200,000$ km. per sec.; but the speed of light relative to an observer travelling with it is still $300,000$ km. per sec. It reminds us of the mathematicians' transfinite number Aleph; you can subtract any number you like from it, and it still remains the same.

The velocity of light plays a conspicuous part in the relativity theory, and it is of importance to understand what is the property associated with it which makes it fundamental. The fact that the velocity of light is the same for all observers is a consequence rather than a cause of its pre-eminent character. Our first introduction of it, for the purpose of coordinating units of length and time, was merely conventional with a view to simplifying the algebraic expressions. Subsequently, considerable use has been made of the fact that nothing is known in physics which travels with greater speed, so that in practice our determinations of simultaneity depend on signals transmitted with this speed. If some new kind of ray with a higher speed were discovered, it would perhaps tend to displace light-signals and light-velocity in this part of the work, time-reckoning being modified to correspond; on the other hand, this would lead to greater complexity in the formulae, because the FitzGerald contraction which affects space-measurement depends on light-velocity. But the chief importance of the velocity of light is that no material body can exceed this velocity. This gives a general physical distinction between paths which are time-like and space-like, respectively—those which can be traversed by matter, and those which cannot. The material structure of the four-dimensional world is fibrous, with the threads all running along time-like tracks; it is a tangled warp without a woof. Hence, even if the discovery of a new ray led us to modify the reckoning of time and space, it would still be necessary in the study of material systems to preserve the *present* absolute distinction of time-like and space-like intervals, under a new name if necessary.

It may be asked whether it is possible for anything to have a speed greater than the velocity of light. Certainly matter cannot attain a greater speed; but there might be other things in nature which could. "Mr Speaker," said Sir Boyle Roche, "not being a bird, I could not be in two places at the same time." Any entity with a speed greater than light would have the peculiarity of Sir Boyle Roche's bird. It can scarcely be said to be a self-contradictory property to be in two places at the same time any more than for an object to be at two times in the same place. The perplexities of the quantum theory of energy sometimes seem to suggest that the possibility ought not to be overlooked; but, on the whole, the evidence seems to be against the existence of anything moving with a speed beyond that of light.

The standpoint of relativity and the principle of relativity are quite independent of any views as to the constitution of matter or light. Hitherto our only reference to electrical theory has been in connection with Larmor and Lorentz's explanation of the FitzGerald contraction; but now from the discussion of the four-dimensional world, we have found a more general explanation of the change of length. The case for the electrical theory of matter is actually weakened, because many experimental effects formerly thought to depend on the peculiar properties of electrical forces are now found to be perfectly general consequences of the relativity of observational knowledge.

Whilst the evidence for the electrical theory of matter is not so conclusive, as at one time appeared, the theory may be accepted without serious misgivings. To postulate two entities, matter and electric charges, when one will suffice is an arbitrary hypothesis, unjustifiable in our present state of knowledge. The great contribution of the electrical theory to this subject is a precise explanation of the property of inertia. It was shown theoretically by J. J. Thomson that if a charged conductor is to be moved or stopped, additional effort will be necessary simply on account of the charge. The conductor has to carry its electric field with it, and force is needed to set the field moving. This property is called inertia, and it is measured by *mass*. If, keeping the charge constant, the size of the conductor is diminished, this inertia increases. Since the smallest separable particles of matter are found by experiment to be very minute and to carry charges, the suggestion arises that these charges may be responsible for the whole of the inertia detected in matter. The explanation is sufficient; and there seems no reason to doubt that all inertia is of this electrical kind.

When the calculations are extended to charges moving with high velocities, it is found that the electrical inertia is not strictly constant but depends on the speed; in all cases the variation is summed up in the statement that the inertia is simply proportional to the total energy of the electromagnetic field. We can say if we like that the mass of a charged particle at rest belongs to its electrostatic energy; when the charge is set in motion, kinetic energy is added, and this kinetic energy also has mass. Hence it appears that mass (inertia) and energy are essentially the same thing, or, at the most, two aspects of the same thing. It must be remembered that on this view the greater part of the mass of matter is due to concealed energy, which is not as yet releasable.

The question whether electrical energy not bound to electric charges has mass, is answered in the affirmative in the case of light. Light has mass. Presumably also gravitational energy has mass; or, if not, mass will be created when, as often happens, gravitational energy is converted into kinetic energy. The mass of the whole (negative) gravitational energy of the earth is of the order *minus* a billion tons.

The theoretical increase of the mass of an electron with speed has been confirmed experimentally, the agreement with calculation being perfect if the electron undergoes the FitzGerald contraction by its motion. This has been held to indicate that the electron cannot have any inertia other than that due to the electromagnetic field carried with it. But the conclusion (though probable enough) is not a fair inference; because these results, obtained by special calculation for electrical inertia, are found to be predicted by the theory of relativity for any kind of inertia. This will be shown in Chapter 9. The factor giving the increase of mass with speed is the same as that which affects length and time. Thus if a rod moves at such a speed that its length is halved, its mass will be doubled. Its density will be increased four-fold, since it is both heavier and less in volume.

We have thought it necessary to include this brief summary of the electrical theory of matter and mass, because, although it is not required by the relativity theory, it is so universally accepted in physics that we can scarcely ignore it. Later on we shall reach in a more general way the identification of mass with energy and the variation of mass with speed; but, since the experimental measurement of inertia involves the study of a body in non-uniform motion, it is not possible to enter on a satisfactory discussion of mass until the more general theory of relativity for non-uniform motion has been developed.

4 FIELDS OF FORCE

For whenever bodies fall through water and thin air, they must quicken their descents in proportion to their weights, because the body of water and subtle nature of air cannot retard everything in equal degree, but more readily give way overpowered by the heavier; on the other hand empty void cannot offer resistance to anything in any direction at any time, but must, as its nature craves, continually give way; and for this reason all things must be moved and borne along with equal velocities though of unequal weights through the unresisting void.

LUCRETIUS, *De Natura Rerum.*

THE primary conception of force is associated with the muscular sensation felt when we make an effort to cause or prevent the motion of matter. Similar effects on the motion of matter can be caused by non-living agency, and these also are regarded as due to forces. As is well known, the scientific measure of a force is the momentum that it communicates to a body in given time. There is nothing very abstract about a force transmitted by material contact; modern physics shows that the momentum is communicated by a process of molecular bombardment. We can visualise the mechanism, and see the molecules carrying the motion in small parcels across the boundary into the body that is being acted on. Force is no mysterious agency; it is merely a convenient summary of this flow of motion, which we can trace continuously if we take the trouble. It is true that the difficulties are only set back a stage, and the exact mode by which the momentum is redistributed during a molecular collision is not yet understood; but, so far as it goes, this analysis gives a clear idea of the transmission of motion by ordinary forces.

But even in elementary mechanics an important natural force appears, which does not seem to operate in this manner. Gravitation is not resolvable into a succession of molecular blows. A massive body, such as the earth, seems to be surrounded by a field of latent force, ready, if another body enters the field, to become active, and transmit motion. One usually thinks of this influence as existing in the space round the earth even when there is no test-body to be affected, and in a rather vague way it is suspected to be some state of strain or other condition of an unperceived medium.

Although gravitation has been recognised for thousands of years, and its laws were formulated with sufficient accuracy for almost all purposes more than 200 years ago, it cannot be said that much progress has been made in explaining the nature or mechanism of this influence. It is said that more than 200 theories of gravitation have been put forward; but the most plausible of these have all had the defect that they lead nowhere and admit of no experimental test. Many of them would nowadays be dismissed as too materialistic for our taste—filling space with the hum of machinery—a procedure curiously popular in the nineteenth century. Few would survive the recent discovery that gravitation acts not only on the molecules of matter, but on the undulations of light.

The nature of gravitation has seemed very mysterious, yet it is a remarkable fact that in a limited region it is possible to create an artificial field of force which imitates a natural gravitational field so exactly that, so far as experiments have yet gone, no one can tell the difference. Those who seek for an explanation of gravitation naturally aim to find a model which will reproduce its effects; but no one before Einstein seems to have thought of finding the clue in these artificial fields, familiar as they are.

When a lift starts to move upwards the occupants feel a characteristic sensation, which is actually identical with a sensation of increased weight. The feeling disappears as soon as the motion becomes uniform; it is associated only with the change of motion of the lift, that is to say, the acceleration. Increased weight is not only a matter of sensation; it is shown by any physical experiments that can be performed. The usual laboratory determination of the value of gravity by Atwood's machine would, if carried out inside the accelerated lift, give a higher value. A spring-balance would record higher weights. Projectiles would follow the usual laws of motion but with a higher value of gravity. In fact, the upward acceleration of the lift is in its mechanical effects exactly similar to an additional gravitational field superimposed on that normally present.

Perhaps the equivalence is most easily seen when we produce in this manner an artificial field which just neutralises the earth's field of gravitation. Jules Verne's book *Round the Moon* tells the story of three men in a projectile shot from a cannon into space. The author enlarges on their amusing experiences when their weight vanished altogether at the neutral point, where the attraction of the earth and moon balance one another. As a matter of fact they would not have had any feeling of weight at any time during their journey after they left the earth's atmosphere. The projectile was responding freely to the pull of gravity, and so were its occupants. When an occupant let go of a plate, the plate could not "fall" any more than it was doing already, and so it must remain poised.

It will be seen that the sensation of weight is not felt when we are free to respond to the force of gravitation; it is only felt when something interferes to prevent our falling. It is primarily the floor or the chair which causes the sensation of weight by checking the fall. It seems literally true to say that we never feel the force of the earth's gravitation; what we do feel is the bombardment of the soles of our boots by the molecules of the ground, and the consequent impulses spreading upwards through the body. This point is of some importance, since the idea of the force of gravitation as something which can be felt, predisposes us to a materialistic view of its nature.

Another example of an artificial field of force is the centrifugal force of the earth's rotation. In most books of Physical Constants will be found a table of the values of "g," the acceleration due to gravity, at different latitudes. But the numbers given do not relate to gravity alone; they are the resultant of gravity and the centrifugal force of the earth's rotation. These are so much alike in their effects that for practical purposes physicists have not thought it worth while to distinguish them.

Similar artificial fields are produced when an aeroplane changes its course or speed; and one of the difficulties of navigation is the impossibility of discriminating between these and the true gravitation of the earth with which they combine. One usually finds that the practical aviator requires little persuasion of the relativity of force.

To find a unifying idea as to the origin of these artificial fields of force, we must return to the four-dimensional world of space-time. The observer is progressing along a certain track in this world. Now his course need not necessarily be straight. It must be remembered that straight in the four-dimensional world means something more than straight in space; it implies also uniform velocity, since the velocity determines the inclination of the track to the time-axis.

The observer in the accelerated lift travels upwards in a straight line, say 1 foot in the first second, 4 feet in two seconds, 9 feet in three seconds, and so on. If we plot these points as x and t on a diagram we obtain a curved track. Presently the speed of the lift becomes uniform and the track in the diagram becomes straight. So long as the track is curved (accelerated motion) a field of force is perceived; it disappears when the track becomes straight (uniform motion).

Again the observer on the earth is carried round in a circle once a day by the earth's rotation; allowing for steady progress through time, the track in four dimensions is a spiral. For an observer at the north pole the track is straight, and there the centrifugal force is zero.

Clearly the artificial field of force is associated with curvature of track, and we can lay down the following rule:—

Whenever the observer's track through the four-dimensional world is curved he perceives an artificial field of force.

The field of force is not only perceived by the observer in his sensations, but reveals itself in his physical measures. It should be understood, however, that the curvature of track must not have been otherwise allowed for. Naturally if the observer in the lift recognises that his measures are affected by his own acceleration and applies the appropriate corrections, the artificial force will be removed by the process. It only exists if he is unaware of, or does not choose to consider, his acceleration.

The centrifugal force is often called "unreal." From the point of view of an observer who does not rotate with the earth, there is no centrifugal force; it only arises for the terrestrial observer who is too lazy to make other allowance for the effects of the earth's rotation. It is commonly thought that this "unreality" quite differentiates it from a "real" force like gravity; but if we try to find the grounds of this distinction they evade us. The centrifugal force is made to disappear if we choose a suitable standard observer not rotating with the earth; the gravitational force was made to disappear when we chose as standard observer an occupant of Jules Verne's falling projectile. If the possibility of annulling a field of force by choosing a suitable standard observer is a test of unreality, then gravitation is equally unreal with centrifugal force.

It may be urged that we have not stated the case quite fairly. When we choose the non-rotating observer the centrifugal force disappears completely and everywhere. When we choose the occupant of the falling projectile, gravitation disappears in his immediate neighbourhood; but he would notice that, although unsupported objects round him experienced no acceleration relative to him, objects on the other side of the earth would fall towards him. So far from getting rid of the field of force, he has merely removed it from his own surroundings, and piled it up elsewhere. Thus gravitation is removable locally, but centrifugal force can be removed everywhere. The fallacy of this argument is that it speaks as though gravitation and centrifugal force were distinguishable experimentally. It presupposes the distinction that we are challenging. Looking simply at the resultant of gravitation and centrifugal force, which is all that can be observed, neither observer can get rid of the resultant force at all parts of space. Each has to be content with leaving a residuum. The non-rotating observer claims that he has got rid of all the unreal part, leaving a remainder (the usual gravitational field) which he regards as really existing. We see no justification for this claim, which might equally well be made by Jules Verne's observer.

It is not denied that the separation of centrifugal and gravitational force generally adopted has many advantages for mathematical calculation. If it were not so, it could not have endured so long. But it is a mathematical separation only, without physical basis; and it often happens that the separation of a mathematical expression into two terms of distinct nature, though useful for elementary work, becomes vitiated for more accurate work by the occurrence of minute cross-terms which have to be taken into account.

Newtonian mechanics proceeds on the supposition that there is some super-observer. If *he* feels a field of force, then that force really exists. Lesser beings, such as the occupants of the falling projectile, have other ideas, but they are the victims of illusion. It is to this super-observer that the mathematician appeals when he starts a dynamical investigation with the words "Take unaccelerated rectangular axes, Ox, Oy, Oz ...." Unaccelerated rectangular axes are the measuring-appliances of the super-observer.

It is quite possible that there might be a super-observer, whose views have a natural right to be regarded as the truest, or at least the simplest. A society of learned fishes would probably agree that phenomena were best described from the point of view of a fish at rest in the ocean. But relativity mechanics finds that there is no evidence that the circumstances of any observer can be such as to make his views pre-eminent. All are on an equality. Consider an observer A in a projectile falling freely to the earth, and an observer B in space out of range of any gravitational attraction. Neither A nor B feel any field of force in their neighbourhood. Yet in Newtonian mechanics an artificial distinction is drawn between their circumstances; B is in no field of force at all, but A is really in a field of force, only its effects are neutralised by his acceleration. But what is this acceleration of A? Primarily it is an acceleration relative to the earth; but then that can equally well be described as an acceleration of the earth relative to A, and it is not fair to regard it as something located with A. Its importance in Newtonian philosophy is that it is an acceleration relative to what we have called the super-observer. This potentate has drawn planes and lines partitioning space, as space appears to him. I fear that the time has come for his abdication.

Suppose the whole system of the stars were falling freely under the uniform gravitation of some vast external mass, like a drop of rain falling to the ground. Would this make any difference to phenomena? None at all. There would be a gravitational field; but the consequent acceleration of the observer and his landmarks would produce a field of force annulling it. Who then shall say what is absolute acceleration?

We shall accordingly give up the attempt to separate artificial fields of force and natural gravitational fields; and call the whole measured field of force the gravitational field, generalising the expression. This field is not absolute, but always requires that some observer should be specified.

It may avoid some mystification if we state at once that there are certain intricacies in the gravitational influence radiating from heavy matter which are distinctive. A theory which did not admit this would run counter to common sense. What our argument has shown is that the characteristic symptom in a region in the neighbourhood of matter is not the field of force; it must be something more intricate. In due course we shall have to explain the nature of this more complex effect of matter on the condition of the world.

Our previous rule, that the observer perceives an *artificial* field of force when he deviates from a straight track, must now be superseded. We need rather a rule determining when he perceives a field of force of any kind. Indeed the original rule has become meaningless, because a straight track is no longer an absolute conception. Uniform motion in a straight line is not the same for an observer rotating with the earth as for a non-rotating observer who takes into account the sinuosity of the rotation. We have decided that these two observers are on the same footing and their judgments merit the same respect. A straight-line in space-time is accordingly not an absolute conception, but is only defined relative to some observer.

Now we have seen that so long as the observer and his measuring-appliances are unconstrained (falling freely) the field of force immediately round him vanishes. It is only when he is deflected from his proper track that he finds himself in the midst of a field of force. Leaving on one side the question of the motion of electrically charged bodies, which must be reserved for more profound treatment, the observer can only leave his proper track if he is being disturbed by material impacts, e.g. the molecules of the ground bombarding the soles of his boots. We may say then that a body does not leave its natural track without visible cause; and any field of force round an observer is the result of his leaving his natural track by such cause. There is nothing mysterious about this field of force; it is merely the reflection in the phenomena of the observer's disturbance; just as the flight of the houses and hedgerows past our railway-carriage is the reflection of our motion with the train.

Our attention is thus directed to the natural tracks of unconstrained bodies, which appear to be marked out in some absolute way in the four-dimensional world. There is no question of an observer here; the body takes the same course in the world whoever is watching it. Different observers will describe the track as straight, parabolical, or sinuous, but it is the same absolute locus.

Now we cannot pretend to predict without reference to experiment the laws determining the nature of these tracks; but we can examine whether our knowledge of the four-dimensional world is already sufficient to specify definite tracks of this kind, or whether it will be necessary to introduce new hypothetical factors. It will be found that it is already sufficient. So far we have had to deal with only one quantity which is independent of the observer and has therefore an absolute significance in the world, namely the *interval* between two events in space and time. Let us choose two fairly distant events P_1 and P_2. These can be joined by a variety of tracks, and the interval-length from P_1 to P_2 along any track can be measured. In order to make sure that the interval-length is actually being measured along the selected track, the method is to take a large number of intermediate points on the track, measure the interval corresponding to each subdivision, and take the sum. It is virtually the same process as measuring the length of a twisty road on a map with a piece of cotton. The interval-length along a particular track is thus something which can be measured absolutely, since all observers agree as to the measurement of the interval for each subdivision. It follows that all observers will agree as to which track (if any) is the shortest track between the two points, judged in terms of interval-length.

This gives a means of defining certain tracks in space-time as having an absolute significance, and we proceed tentatively to identify them with the natural tracks taken by freely moving particles.

In one respect we have been caught napping. Dr A. A. Robb has pointed out the curious fact that it is not the shortest track, but the longest track, which is unique[1]. There are any number of tracks from P_1 to P_2 of zero interval-length; there is just one which has maximum length. This is because of the peculiar geometry which the minus sign of $(t_2 - t_1)^2$ introduces. For instance, it will be seen from equation (1), 38, that when

$$(x_2 - x_1)^2 + (y_2 - y_1)^2 + (z_2 - z_1)^2 = (t_2 - t_1)^2,$$

that is to say when the resultant distance travelled in space is equal to the distance travelled in time, then s is zero. This happens when the velocity is unity—the velocity of light. To get from P_1 to P_2 by a path of no interval-length, we must simply keep on travelling with the velocity of light, cruising round if necessary, until the moment comes to turn up at P_2. On the other hand there is evidently an upper limit to the interval-length of the track, because each portion of s is always less than the corresponding portion of $(t_2 - t_1)$, and s can never exceed $t_2 - t_1$.

There is a physical interpretation of interval-length along the path of a particle which helps to give a more tangible idea of its meaning. It is the time as perceived by an observer, or measured by a clock, carried on the particle. This is called the proper-time; and, of course, it will not in general agree with the time-reckoning of the independent onlooker who is supposed to be watching the whole proceedings. To prove this, we notice from equation (1) that if $x_2 = x_1, y_2 = y_1$ and $z_2 = z_1$, then $s = t_2 - t_1$. The condition $x_2 = x_1$, etc. means that the particle must remain stationary relative to the observer who is measuring x, y, z, t. To

[1] It is here assumed that P_2 is in the future of P_1 so that it is possible for a particle to travel from P_1 to P_2. If P_1 and P_2 are situated like O and P' in Fig. 3, the interval-length is imaginary, and the *shortest* track is unique.

secure this we mount our observer on the particle and then the interval-length s will be $t_2 - t_1$, which is the time elapsed according to his clock.

We can use proper-time as generally equivalent to interval-length; but it must be admitted that the term is not very logical unless the track in question is a natural track. For any other track, the drawback to defining the interval-length as the time measured by a clock which follows the track, is that no clock could follow the track without violating the laws of nature. We may force it into the track by continually hitting it; but that treatment may not be good for its time-keeping qualities. The original definition by equation (1) is the more general definition.

We are now able to state formally our proposed law of motion—Every particle moves so as to take the track of greatest interval-length between two events, except in so far as it is disturbed by impacts of other particles or electrical forces.

This cannot be construed into a truism like Newton's first law of motion. The reservation is not an undefined agency like force, whose meaning can be extended to cover any breakdown of the law. We reserve only direct material impacts and electromagnetic causes, the latter being outside our present field of discussion.

Consider, for example, two events in space-time, viz. the position of the earth at the present moment, and its position a hundred years ago. Call these events P_2 and P_1. In the interim the earth (being undisturbed by impacts) has moved so as to take the longest track from P_1 to P_2—or, if we prefer, so as to take the longest possible proper-time over the journey. In the weird geometry of the part of space-time through which it passes (a geometry which is no doubt associated in some way with our perception of the existence of a massive body, the sun) this longest track is a spiral—a circle in space, drawn out into a spiral by continuous displacement in time. Any other course would have had shorter interval-length.

In this way the study of fields of force is reduced to a study of geometry. To a certain extent this is a retrograde step; we adopt Kepler's description of the sun's gravitational field instead of Newton's. The field of force is completely described if the tracks through space and time of particles projected in every possible way are prescribed. But we go back in order to go forward in a new direction. To express this unmanageable mass of detail in a unified way, a world-geometry is found in which the tracks of greatest length are the actual tracks of the particles. It only remains to express the laws of this geometry in a concise form. The change from a mechanical to a geometrical theory of fields of force is not so fundamental a change as might be supposed. If we are now reducing mechanics to a branch of natural geometry, we have to remember that natural geometry is equally a branch of mechanics, since it is concerned with the behaviour of material measuring-appliances.

Reference has been made to weird geometry. There is no help for it, if the longest track can be a spiral like that known to be described by the earth. Non-Euclidean geometry is necessary. In Euclidean geometry the shortest track is always a straight line; and the slight modification of Euclidean geometry described in Chapter 3 is found to give a straight line as the longest track. The status of non-Euclidean geometry has already been thrashed out in the Prologue; and there seems to be no reason whatever for preferring Euclid's geometry unless observations decide in its favour. equation (1), 38, is the expression of the Euclidean (or semi-Euclidean) geometry we have hitherto adopted; we shall have to modify it, if we adopt non-Euclidean geometry.

But the point arises that the geometry arrived at in Chapter 3 was not arbitrary. It was the synthesis of measures made with clocks and scales, by observers with all kinds of uniform motion relative to one another; we cannot modify it arbitrarily to fit the behaviour of moving particles like the earth. Now, if the worst came to the worst, and we could not reconcile a geometry based on measures with clocks and scales and a geometry based on the natural tracks of moving particles—if we had to select one or the other and keep to it—I think we ought to prefer to use the geometry based on the tracks of moving particles. The free motion of a particle is an example of the simplest possible kind of phenomenon; it is unanalysable; whereas, what the readings of any kind of clock record, what the extension of a material rod denotes, may evidently be complicated phenomena involving the secrets of molecular constitution. Each geometry would be right in its own sphere; but the geometry of moving particles would be the more fundamental study. But it turns out that there is probably no need to make the choice; clocks, scales, moving particles, light-pulses, give the same geometry. This might perhaps be expected since a clock must comprise moving particles of some kind.

A formula, such as equation (1), based on experiment can only be verified to a certain degree of approximation. Within certain limits it will be possible to introduce modifications. Now it turns out that the free motion of a particle is a much more sensitive way of exploring space-time, than any practicable measures with scales and clocks. If then we employ our accurate knowledge of the motion of particles to correct the formula, we shall find that the changes introduced are so small that they are inappreciable in any practical measures with scales and clocks. There is only one case where a possible detection of the modification is indicated; this refers to the behaviour of a clock on the surface of the sun, but the experiment is one of great difficulty and no conclusive answer has been given. We conclude then that the geometry of space and time based on the motions of particles is accordant with the geometry based on the cruder observations with clocks and scales; but if subsequent experiment should reveal a discrepancy, we shall adhere to the moving particle on account of its greater simplicity.

The proposed modification can be regarded from another point of view. equation (1) is the synthesis of the experiences of all observers in uniform motion. But uniform motion means that their four-dimensional tracks are straight lines. We must suppose that the observers were moving in their natural tracks; for, if not, they experienced fields of force, and presumably allowed for these in their calculations, so that reduction was made to the natural tracks. If then equation (1) shows that the natural tracks are straight lines, we are merely getting out of the equation that which we originally put into it.

The formula needs generalising in another way. Suppose there is a region of space-time where, for some observer, the natural tracks are all straight lines and equation (1) holds rigorously. For another (accelerated) observer the tracks will be curved, and the equation will not hold. At the best it is of a form which can only hold good for specially selected observers.

Although it has become necessary to throw our formula into the melting-pot, that does not create any difficulty in measuring the interval. Without going into technical details, it may be pointed out that the innovations arise solely from the introduction of gravitational fields of force into our scheme. When there is no force, the tracks of all particles are straight lines as our previous geometry requires. In any small region we can choose an observer (falling freely) for whom the force vanishes, and accordingly the original formula holds good. Thus it is only necessary to modify our rule for determining the interval by two provisos (1) that the interval measured must be small, (2) that the scales and clocks used for measuring it must be falling freely. The second proviso is natural, because, if we do not leave our apparatus to fall freely, we must allow for the strain that it undergoes. The first is not a serious disadvantage, because a larger interval can be split up into a number of small intervals and the parts measured separately. In mathematical problems the same device is met with under the name of integration. To emphasise that the formula is strictly true only for infinitesimal intervals, it is written with a new notation

$$ds^2 = -\,dx^2 - dy^2 - dz^2 + dt^2 \tag{2}$$

where dx stands for the small difference $x_2 - x_1$, etc.

The condition that the measuring appliances must not be subjected to a field of force is illustrated by Ehrenfest's paradox. Consider a wheel revolving rapidly. Each portion of the circumference is moving in the direction of its length, and might be expected to undergo the FitzGerald contraction due to its velocity; each portion of a radius is moving transversely and would therefore have no longitudinal contraction. It looks as though the rim of the wheel should contract and the spokes remain the same length, when the wheel is set revolving. The conclusion is absurd, for a revolving wheel has no tendency to buckle—which would be the only way of reconciling these conditions. The point which the argument has overlooked is that the results here appealed to apply to unconstrained bodies, which have no acceleration relative to the natural tracks in space. Each portion of the rim of the wheel has a radial acceleration, and this affects its extensional properties. When accelerations as well as velocities occur a more far-reaching theory is needed to determine the changes of length.

To sum up—the interval between two (near) events is something quantitative which has an absolute significance in nature. The track between two (distant) events which has the longest interval-length must therefore have an absolute significance. Such tracks are called *geodesics.* Geodesics can be traced practically, because they are the tracks of particles undisturbed by material impacts. By the practical tracing of these geodesics we have the best means of studying the character of the natural geometry of the world. An auxiliary method is by scales and clocks, which, it is believed, when unconstrained, measure a small interval according to formula (2).

The identity of the two methods of exploring the geometry of the world is connected with a principle which must now be enunciated definitely. We have said that no experiments have been able to detect a difference between a gravitational field and an artificial field of force such as the centrifugal force. This is not quite the same thing as saying that it has been proved that there is no difference. It is well to be explicit when a positive generalisation is made from negative experimental evidence. The generalisation which it is proposed to adopt is known as the Principle of Equivalence.

A gravitational field of force is precisely equivalent to an artificial field of force, so that in any small region it is impossible by any conceivable experiment to distinguish between them.

In other words, force is purely relative.

5 KINDS OF SPACE

The danger of asserting dogmatically that an axiom based on the experience of a limited region holds universally will now be to some extent apparent to the reader. It may lead us to entirely overlook, or when suggested at once reject, a possible explanation of phenomena. The hypothesis that space is not homaloidal [flat], and again that its geometrical character may change with the time, may or may not be destined to play a great part in the physics of the future; yet we cannot refuse to consider them as possible explanations of physical phenomena, because they may be opposed to the popular dogmatic belief in the universality of certain geometrical axioms—a belief which has risen from centuries of indiscriminating worship of the genius of Euclid.

W. K. CLIFFORD(and K. PEARSON), *Common Sense of the Exact Sciences.*

ON any surface it requires two independent numbers or "coordinates" to specify the position of a point. For this reason a surface, whether flat or curved, is called a two-dimensional space. Points in three-dimensional space require three, and in four-dimensional space-time four numbers or coordinates.

To locate a point on a surface by two numbers, we divide the surface into meshes by any two systems of lines which cross one another. Attaching consecutive numbers to the lines, or better to the channels between them, one number from each system will identify a particular mesh; and if the subdivision is sufficiently fine any point can be specified in this way with all the accuracy needed. This method is used, for example, in the Post Office Directory of London for giving the location of streets on the map. The point $(4, 2)$ will be a point in the mesh where channel No. 4 of the first system crosses channel No. 2 of the second. If this indication is not sufficiently accurate, we must divide channel No. 4 into ten parts numbered 4.0, 4.1 etc. The subdivision must be continued until the meshes are so small that all points in one mesh can be considered identical within the limits of experimental detection.

The diagrams, Figs. 10, 11, 12, illustrate three of the many kinds of mesh-systems commonly used on a flat surface.

If we speak of the properties of the triangle formed by the points $(1, 2)$, $(3, 0)$, $(4, 4)$, we shall be at once asked, What mesh-system are you using? No one can form a picture of the triangle until that information has been given. But if we speak of the properties of a triangle whose sides are of lengths 2, 3, 4 inches, anyone with a graduated scale can draw the triangle, and follow our discussion of its properties. The distance between two points can be stated without referring to any mesh-system. For this reason, if we use a mesh-system, it is important to find formulae connecting the absolute distance with the particular system that is being used.

In the more complicated kinds of mesh-systems it makes a great simplification if we content ourselves with the formulae for very short distances. The mathematician then finds no difficulty in extending the results to long distances by the process called integration. We write ds for the distance between two points close

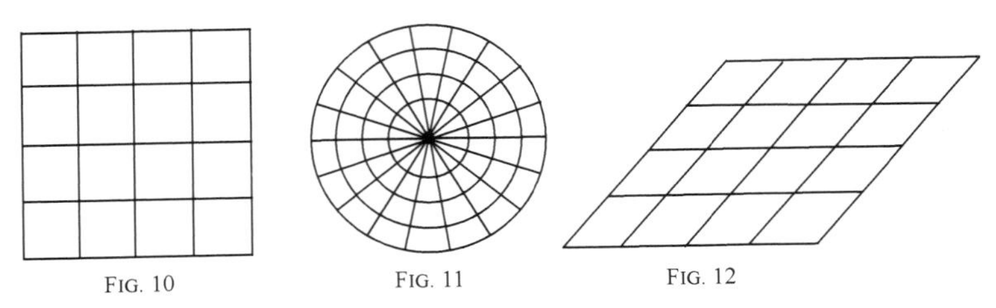

FIG. 10 FIG. 11 FIG. 12

together, x_1 and x_2 for the two numbers specifying the location of one of them, dx_1 and dx_2 for the small differences of these numbers in passing from the first point to the second. But in using one of the particular mesh-systems illustrated in the diagrams, we usually replace x_1, x_2 by particular symbols sanctioned by custom, viz. (x_1, x_2) becomes (x, y), (r, θ), (ξ, η) for Figs. 10, 11, 12, respectively.

The formulae, found by geometry, are:

For rectangular coordinates (x, y), Fig. 10,

$$ds^2 = dx^2 + dy^2.$$

For polar coordinates (r, θ), Fig. 11,

$$ds^2 = dr^2 + r^2\, d\theta^2.$$

For oblique coordinates (ξ, η), Fig. 12,

$$ds^2 = d\xi^2 - 2\kappa\, d\xi d\eta + d\eta^2,$$

where κ is the cosine of the angle between the lines of partition.

As an example of a mesh-system on a curved surface, we may take the lines of latitude and longitude on a sphere.

For latitude and longitude (β, λ)

$$ds^2 = d\beta^2 + \cos^2 \beta\, d\lambda^2.$$

These expressions form a test, and in fact the only possible test, of the kind of coordinates we are using. It may perhaps seem inconceivable that an observer should for an instant be in doubt whether he was using the mesh-system of Fig. 10 or Fig. 11. He sees at a glance that Fig. 11 is not what he would call a rectangular mesh-system. But in that glance, he makes measures with his eye, that is to say he determines ds for pairs of points, and he notices how these values are related to the number of intervening channels. In fact he is testing which formula for ds will fit. For centuries man was in doubt whether the earth was flat or round—whether he was using plane rectangular coordinates or some kind of spherical coordinates. In some cases an observer adopts his mesh-system blindly and long afterwards discovers by accurate measures that ds does not fit the formula he assumed—that his mesh-system is not exactly of the nature he supposed it was. In other cases he deliberately sets himself to plan out a mesh-system of a particular variety, say rectangular coordinates; he constructs right angles and rules parallel lines; but these constructions are all measurements of the way the x-channels and y-channels ought to go, and the rules of construction reduce to a formula connecting his measures ds with x and y.

The use of special symbols for the coordinates, varying according to the kind of mesh-system used, thus anticipates a knowledge which is really derived from the form of the formulae. In order not to give away the secret prematurely, it will be better to use the symbols x_1, x_2 in all cases. The four kinds of coordinates already considered then give respectively the relations,

$$\begin{aligned} ds^2 &= dx_1{}^2 + dx_2{}^2 && \text{(rectangular)}, \\ ds^2 &= dx_1{}^2 + x_1{}^2\, dx_2{}^2 && \text{(polar)}, \\ ds^2 &= dx_1{}^2 - 2\kappa\, dx_1 dx_2 + dx_2{}^2 && \text{(oblique)}, \\ ds^2 &= dx_1{}^2 + \cos^2 x_1\, dx_2{}^2 && \text{(latitude and longitude)}. \end{aligned}$$

If we have any mesh-system and want to know its nature, we must make a number of measures of the length ds between adjacent points (x_1, x_2) and $(x_1 + dx_1, x_2 + dx_2)$ and test which formula fits. If, for example, we then find that ds^2 is always equal to $dx_1{}^2 + x_1{}^2\, dx_2{}^2$, we know that our mesh-system is like that in Fig. 11, x_1 and x_2 being the numbers usually denoted by the polar coordinates r, θ. The statement that polar coordinates are being used is unnecessary, because it adds nothing to our knowledge which is not already contained in the formula. It is merely a matter of giving a name; but, of course, the name calls to our minds a number of familiar properties which otherwise might not occur to us.

For instance, it is characteristic of the polar coordinate system that there is only one point for which x_1 (or r) is equal to 0, whereas in the other systems $x_1 = 0$ gives a line of points. This is at once apparent from the formula; for if we have two points for which $x_1 = 0$ and $x_1 + dx_1 = 0$, respectively, then

$$dx_1{}^2 + x_1{}^2\, dx_2{}^2 = 0.$$

The distance ds between the two points vanishes, and accordingly they must be the same point.

The examples given can all be summed up in one general expression

$$ds^2 = g_{11}\, dx_1{}^2 + 2g_{12}\, dx_1 dx_2 + g_{22}\, dx_2{}^2,$$

where g_{11}, g_{12}, g_{22} may be constants or functions of x_1 and x_2. For instance, in the fourth example their values are 1, 0, $\cos^2 x_1$. It is found that all possible mesh-systems lead to values of ds^2 which can be included in an expression of this general form; so that mesh-systems are distinguished by three functions of position g_{11}, g_{12}, g_{22} which can be determined by making physical measurements. These three quantities are sometimes called potentials.

We now come to a point of far-reaching importance. The formula for ds^2 teaches us not only the character of the mesh-system, but the nature of our two-dimensional space, which is independent of any mesh-system. If ds^2 satisfies any one of the first three formulae, then the space is like a flat surface; if it satisfies the last formula, then the space is a surface curved like a sphere. Try how you will, you cannot draw a mesh-system on a flat (Euclidean) surface which agrees with the fourth formula.

If a being limited to a two-dimensional world finds that his measures agree with the first formula, he can make them agree with the second or third formulae by drawing the meshes differently. But to obtain the fourth formula he must be translated to a different world altogether.

We thus see that there are different kinds of two-dimensional space, betrayed by different metrical properties. They are naturally visualised as different surfaces in Euclidean space of three dimensions. This picture is helpful in some ways, but perhaps misleading in others. The metrical relations on a plane sheet of paper are not altered when the paper is rolled into a cylinder—the measures being, of course, confined to the two-dimensional world represented by the paper, and not allowed to take a short cut through space. The formulae apply equally well to a plane surface or a cylindrical surface; and in so far as our picture draws a distinction between a plane and a cylinder, it is misleading. But they do not apply to a sphere, because a plane sheet of paper cannot be wrapped round a sphere. A genuinely two-dimensional being could not be cognisant of the difference between a cylinder[1] and a plane; but a sphere would appear as a different kind of space, and he would recognise the difference by measurement.

Of course there are many kinds of mesh-systems, and many kinds of two-dimensional spaces, besides those illustrated in the four examples. Clearly it is not going to be a simple matter to discriminate the different kinds of spaces by the values of the g's. There is no characteristic, visible to cursory inspection, which suggests why the first three formulae should all belong to the same kind of space, and the fourth to a different one. Mathematical investigation has discovered what is the common link between the first three formulae. The g_{11}, g_{12}, g_{22} satisfy in all three cases a certain differential equation;[2]and whenever this differential equation is satisfied, the same kind of space occurs.

No doubt it seems a very clumsy way of approaching these intrinsic differences of kinds of space—to introduce potentials which specifically refer to a particular mesh-system, although the mesh-system can have nothing to do with the matter. It is worrying not to be able to express the differences of space in a purer form without mixing them up with irrelevant differences of potential. But we have neither the vocabulary nor the imagination for a description of absolute properties as such. All physical knowledge is relative to space and time partitions; and to gain an understanding of the absolute it is necessary to approach it through the relative. The absolute may be defined as a relative which is always the same no matter what it is relative to[3]. Although we think of it as self-existing, we cannot give it a place in our knowledge without setting up some dummy to relate it to. And similarly the absolute differences of space always appear as related to some mesh-system, although the mesh-system is only a dummy and has nothing

[1]One should perhaps rather say a roll, to avoid any question of joining the two edges.

[2]Appendix, Note 4.

[3]Cf. p. 23, where a distinction was drawn between knowledge which does not particularise the observer and knowledge which does not postulate an observer at all.

to do with the problem.

The results for two dimensions can be generalised, and applied to four-dimensional space-time. Distance must be replaced by interval, which it will be remembered, is an absolute quantity, and therefore independent of the mesh-system used. Partitioning space-time by any system of meshes, a mesh being given by the crossing of four channels, we must specify a point in space-time by four coordinate numbers, x_1, x_2, x_3, x_4. By analogy the general formula will be

$$\begin{aligned} ds^2 = {} & g_{11}\,dx_1{}^2 + g_{22}\,dx_2{}^2 + g_{33}\,dx_3{}^2 + g_{44}\,dx_4{}^2 \\ & + 2g_{12}\,dx_1dx_2 + 2g_{13}\,dx_1dx_3 + 2g_{14}\,dx_1dx_4 \\ & + 2g_{23}\,dx_2dx_3 + 2g_{24}\,dx_2dx_4 + 2g_{34}\,dx_3dx_4. \end{aligned} \tag{3}$$

The only difference is that there are now ten g's, or potentials, instead of three, to summarise the metrical properties of the mesh-system. It is convenient in specifying special values of the potentials to arrange them in the standard form

$$\begin{matrix} g_{11} & g_{12} & g_{13} & g_{14} \\ & g_{22} & g_{23} & g_{24} \\ & & g_{33} & g_{34} \\ & & & g_{44} \end{matrix}$$

The space-time already discussed at length in Chapter 3 corresponded to the formula (2), p. 55,

$$ds^2 = -dx^2 - dy^2 - dz^2 + dt^2.$$

Here (x, y, z, t) are the conventional symbols for (x_1, x_2, x_3, x_4) when this special mesh-system is used, viz. rectangular coordinates and time. Comparing with (3) the potentials have the special values

$$\begin{matrix} -1 & 0 & 0 & 0 \\ & -1 & 0 & 0 \\ & & -1 & 0 \\ & & & +1 \end{matrix}$$

These are called the "Galilean values." If the potentials have these values everywhere, space-time may be called "flat," because the geometry is that of a plane surface drawn in Euclidean space of five dimensions. Recollecting what we found for two dimensions, we shall realise that a quite different set of values of the potentials may also belong to flat space-time, because the meshes may be drawn in different ways. We must clearly understand that

(1) The only way of discovering what kind of space-time is being dealt with is from the values of the potentials, which are determined practically by measurements of intervals,

(2) Different values of the potentials do not necessarily indicate different kinds of space-time,

(3) There is some complicated mathematical property common to all values of the potentials which belong to the same space-time, which is not shared by those which belong to a different kind of space-time. This property is expressed by a set of differential equations.

It can now be deduced that the space-time in which we live is not quite flat. If it were, a mesh-system could be drawn for which the g's have the Galilean values, and the geometry with respect to these partitions of space and time would be that discussed in Chapter 3. For that geometry the geodesics, giving the natural tracks of particles, are straight lines.

Thus in flat space-time the law of motion is that (with suitably chosen co-ordinates) every particle moves uniformly in a straight line except when it is disturbed by the impacts of other particles. Clearly this is not true of our world; for example, the planets do not move in straight lines although they do not suffer any impacts. It is true that if we confine attention to a small region like the interior of Jules Verne's projectile, all the tracks become straight lines for an appropriate observer, or, as we generally say, he detects no field of force. It needs a large region to bring out the differences of geometry. That is not surprising, because we cannot expect to tell whether a surface is flat or curved unless we consider a reasonably large portion of it.

According to Newtonian ideas, at a great distance from all matter beyond the reach of any gravitation, particles would all move uniformly in straight lines. Thus at a great distance from all matter space-time tends to become perfectly flat. This can only be checked by experiment to a certain degree of accuracy, and there is some doubt as to whether it is rigorously true. We shall leave this afterthought to Chapter 3, meanwhile assuming with Newton that space-time far enough away from everything is flat, although near matter it is curved. It is this puckering near matter which accounts for its gravitational effects.

Just as we picture different kinds of two-dimensional space as differently curved surfaces in our ordinary space of three-dimensions, so we are now picturing different kinds of four-dimensional space-time as differently curved surfaces in a Euclidean space of *five* dimensions. This is a picture only[4]. The fifth dimension is neither space nor time nor anything that can be perceived; so far as we know, it is nonsense. I should not describe it as a mathematical fiction, because it is of no great advantage in a mathematical treatment. It is even liable to mislead because it draws distinctions, like the distinction between a plane and a roll, which have no meaning. It is, like the notion of a field of force acting in space and time, merely introduced to bolster up Euclidean geometry, when Euclidean geometry has been found inappropriate. The real difference between the various kinds of space-time is that they have different kinds of geometry, involving different properties of the g's. It is no explanation to say that this is because the surfaces are differently curved in a real Euclidean space of five dimensions. We should naturally ask for an explanation why the space of five dimensions is Euclidean; and presumably the answer would be, because it is a plane in a real Euclidean space of six dimensions, and so on *ad infinitum*.

The value of the picture to us is that it enables us to describe important properties with common terms like "pucker" and "curvature" instead of technical terms like "differential invariant." We have, however, to be on our guard, because

[4]A fifth dimension suffices for illustrating the property here considered; but for an exact representation of the geometry of the world, Euclidean space of *ten* dimensions is required. We may well ask whether there is merit in Euclidean geometry sufficient to justify going to such extremes.

analogies based on three-dimensional space do not always apply immediately to many-dimensional space. The writer has keen recollections of a period of much perplexity, when he had not realised that a four-dimensional space with "no curvature" is not the same as a "flat" space! Three-dimensional geometry does not prepare us for these surprises.

Picturing the space-time in the gravitational field round the earth as a pucker, we notice that we cannot locate the pucker at a point; it is "somewhere round" the point. At any special point the pucker can be pressed out flat, and the irregularity runs off somewhere else. That is what the inhabitants of Jules Verne's projectile did; they flattened out the pucker inside the projectile so that they could not detect any field of force there; but this only made things worse somewhere else, and they would find an increased field of force (relative to them) on the other side of the earth.

What determines the existence of the pucker is not the values of the g's at any point, or, what comes to the same thing, the field of force there. It is the way these values link on to those at other points—the gradient of the g's, and more particularly the gradient of the gradient. Or, as has already been said, the kind of space-time is fixed by differential equations.

Thus, although a gravitational field of force is not an absolute thing, and can be imitated or annulled at any point by an acceleration of the observer or a change of his mesh-system, nevertheless the presence of a heavy particle does modify the world around it in an absolute way which cannot be imitated artificially. Gravitational force is relative; but there is this more complex character of gravitational influence which is absolute.

The question must now be put, Can every possible kind of space-time occur in an empty region in nature? Suppose we give the ten potentials perfectly arbitrary values at every point; that will specify the geometry of some mathematically possible space-time. But could that kind of space-time actually occur—by any arrangement of the matter round the region?

The answer is that only certain kinds of space-time can occur in an empty region in nature. The law which determines what kinds can occur is the law of gravitation.

It is indeed clear that, since we have reduced the theory of fields of force to a theory of the geometry of the world, if there is any law governing fields of force (including the gravitational field), that law must be of the nature of a restriction on the possible geometries of the world.

The choice of g's in any special problem is thus arrived at by a three-fold sorting out: (1) many sets of values can be dismissed because they can never occur in nature, (2) others, while possible, do not relate to the kind of space-time present in the problem considered, (3) of those which remain, one set of values relates to the particular mesh-system that has been chosen. We have now to find the law governing the first discrimination. What is the criterion that decides what values of the g's give a kind of space-time possible in nature?

In solving this problem Einstein had only two clues to guide him.

(1) Since it is a question of whether the *kind of space-time* is possible, the criterion must refer to those properties of the g's which distinguish different kinds of space-time, not to those which distinguish different kinds of mesh-system in

the same space-time. The formulae must therefore not be altered in any way, if we change the mesh-system.

(2) We know that flat space-time *can* occur in nature (at great distances from all gravitating matter). Hence the criterion must be satisfied by any values of the g's belonging to flat space-time.

It is remarkable that these slender clues are sufficient to indicate almost uniquely a particular law. Afterwards the further test must be applied—whether the law is confirmed by observation.

The irrelevance of the mesh-system to the laws of nature is sometimes expressed in a slightly different way. There is one type of observation which, we can scarcely doubt, must be independent of any possible circumstances of the observer, namely a complete coincidence in space and time. The track of a particle through four-dimensional space-time is called its world-line. Now, the world-lines of two particles either intersect or they do not intersect; the standpoint of the observer is not involved. In so far as our knowledge of nature is a knowledge of intersections of world-lines, it is absolute knowledge independent of the observer. If we examine the nature of our observations, distinguishing what is actually seen from what is merely inferred, we find that, at least in all exact measurements, our knowledge is primarily built up of intersections of world-lines of two or more entities, that is to say their coincidences. For example, an electrician states that he has observed a current of 5 milliamperes. This is his inference: his actual observation was a *coincidence* of the image of a wire in his galvanometer with a division of a scale. A meteorologist finds that the temperature of the air is 75; his observation was the *coincidence* of the top of the mercury-thread with division 75 on the scale of his thermometer. It would be extremely clumsy to describe the results of the simplest physical experiment entirely in terms of coincidence. The absolute observation is, whether or not the coincidence exists, not when or where or under what circumstances the coincidence exists; unless we are to resort to relative knowledge, the place, time and other circumstances must in their turn be described by reference to other coincidences. But it seems clear that if we could draw all the world-lines so as to show all the intersections in their proper order, but otherwise arbitrary, this would contain a complete history of the world, and nothing within reach of observation would be omitted.

Let us draw such a picture, and imagine it embedded in a jelly. If we deform the jelly in any way, the intersections will still occur in the same order along each world-line and no additional intersections will be created. The deformed jelly will represent a history of the world, just as accurate as the one originally drawn; there can be no criterion for distinguishing which is the best representation.

Suppose now we introduce space and time-partitions, which we might do by drawing rectangular meshes in both jellies. We have now two ways of locating the world-lines and events in space and time, both on the same absolute footing. But clearly it makes no difference in the result of the location whether we first deform the jelly and then introduce regular meshes, or whether we introduce irregular meshes in the undeformed jelly. And so all mesh-systems are on the same footing.

This account of our observational knowledge of nature shows that there is no *shape* inherent in the absolute world, so that when we insert a mesh-system, it has no shape initially, and a rectangular mesh-system is intrinsically no different from any other mesh-system.

Returning to our two clues, condition (1) makes an extraordinarily clean sweep of laws that might be suggested; among them Newton's law is swept away. The mode of rejection can be seen by an example; it will be sufficient to consider two dimensions. If in one mesh-system (x, y)

$$ds^2 = g_{11}dx^2 + 2g_{12}dxdy + g_{22}dy^2,$$

and in another system (x', y')

$$ds^2 = g'_{11}\,dx'^2 + 2g'_{12}\,dx'dy' + g'_{22}\,dy'^2,$$

the same law must be satisfied if the unaccented letters are throughout replaced by accented letters. Suppose the law $g_{11} = g_{22}$ is suggested. Change the mesh-system by spacing the y-lines twice as far apart, that is to say take $y' = \frac{1}{2}y$, with $x' = x$. Then

$$\begin{aligned} ds^2 &= g_{11}\,dx^2 + 2g_{12}\,dxdy + g_{22}\,dy^2 \\ &= g_{11}\,dx'^2 + 4g_{12}\,dx'dy' + 4g_{22}\,dy'^2, \end{aligned}$$

so that

$$g_{11}' = g_{11}, \qquad g_{22}' = 4g_{22}.$$

And if g_{11} is equal to g_{22}, g'_{11} cannot be equal to g'_{22}.

After a few trials the reader will begin to be surprised that any possible law could survive the test. It seems so easy to defeat any formula that is set up by a simple change of mesh-system. Certainly it is unlikely that anyone would hit on such a law by trial. But there are such laws, composed of exceedingly complicated mathematical expressions. The theory of these is called the "theory of tensors," and had already been worked out by the pure mathematicians Riemann, Christoffel, Ricci, Levi-Civita who, it may be presumed, never dreamt of a physical application for it.

One law of this kind is the condition for flat space-time, which is generally written in the simple, but not very illuminating, form

$$B^{\rho}_{\mu\nu\sigma} = 0. \tag{4}$$

The quantity on the left is called the Riemann-Christoffel tensor, and it is written out in a less abbreviated form in the Appendix[5]. It must be explained that the letters μ, ν, σ, ρ indicate *gaps*, which are to be filled up by any of the numbers 1, 2, 3, 4, chosen at pleasure. (When the expression is written out at length, the gaps are in the suffixes of the x's and g's.) Filling the gaps in different ways, a large number of expressions, B^1_{111}, B^4_{123}, B^1_{432}, etc., are obtained. The equation (4) states that all of these are zero. There are 4^4, or 256, of these expressions altogether, but many of them are repetitions. Only 20 of the equations are really necessary; the others merely say the same thing over again.

---

[5] Appendix, Note 5.

It is clear that the law (4) is not the law of gravitation for which we are seeking, because it is much too drastic. If it were a law of nature, then only flat space-time could exist in nature, and there would be no such thing as gravitation. It is not the general condition, but a special case—when all attracting matter is infinitely remote.

But in finding a general condition, it may be a great help to know a special case. Would it do to select a certain number of the 20 equations to be satisfied generally, leaving the rest to be satisfied only in the special case? Unfortunately the equations hang together; and, unless we take them all, it is found that the condition is not independent of the mesh-system. But there happens to be one way of building up out of the 20 conditions a less stringent set of conditions independent of the mesh-system. Let

$$G_{11} = B^1_{111} + B^2_{112} + B^3_{113} + B^4_{114},$$

and, generally

$$G_{\mu\nu} = B^1_{\mu\nu 1} + B^2_{\mu\nu 2} + B^3_{\mu\nu 3} + B^4_{\mu\nu 4},$$

then the conditions

$$G_{\mu\nu} = 0 \tag{5}$$

will satisfy our requirements for a general law of nature.

This law is independent of the mesh-system, though this can only be proved by elaborate mathematical analysis. Evidently, when all the B's vanish, equation (5) is satisfied; so, when flat space-time occurs, this law of nature is not violated. Further it is not so stringent as the condition for flatness, and admits of the occurrence of a limited variety of non-Euclidean geometries. Rejecting duplicates, it comprises 10 equations; but four of these can be derived from the other six, so that it gives six conditions, which happens to be the number required for a law of gravitation[6].

The suggestion is thus reached that

$$G_{\mu\nu} = 0$$

may be the general law of gravitation. Whether it is so or not can only be settled by experiment. In particular, it must in ordinary cases reduce to something so near the Newtonian law, that the remarkable confirmation of the latter by observation is accounted for. Further it is necessary to examine whether there are any exceptional cases in which the difference between it and Newton's law can be tested. We shall see that these tests are satisfied.

[6]Isolate a region of empty space-time; and suppose that everywhere outside the region the potentials are known. It should then be possible by the law of gravitation to determine the nature of space-time in the region. Ten differential equations together with the boundary-values would suffice to determine the ten potentials throughout the region; but that would determine not only the kind of space-time but the mesh-system, whereas the partitions of the mesh-system can be continued across the region in any arbitrary way. The four sets of partitions give a four-fold arbitrariness; and to admit of this, the number of equations required is reduced to six.

What would have been the position if this suggested law had failed? We might continue the search for other laws satisfying the two conditions laid down; but these would certainly be far more complicated mathematically. I believe too that they would not help much, because practically they would be indistinguishable from the simpler law here suggested—though this has not been demonstrated rigorously. The other alternative is that there is something causing force in nature not comprised in the geometrical scheme hitherto considered, so that force is not purely relative, and Newton's super-observer exists.

Perhaps the best survey of the meaning of our theory can be obtained from the standpoint of a ten-dimensional Euclidean continuum, in which space-time is conceived as a particular four-dimensional surface. It has to be remarked that in ten dimensions there are gradations intermediate between a flat surface and a fully curved surface, which we shall speak of as curved in the "first degree" or "second degree[7]." The distinction is something like that of curves in ordinary space, which may be *curved* like a circle, or *twisted* like a helix; but the analogy is not very close. The full "curvature" of a surface is a single quantity called G, built up out of the various terms $G_{\mu\nu}$ in somewhat the same way as these are built up out of $B^{\rho}_{\mu\nu\sigma}$. The following conclusions can be stated.

If

$$B^{\rho}_{\mu\nu\sigma} = 0 \qquad \text{(20 conditions)}$$

space-time is flat. This is the state of the world at an infinite distance from all matter and all forms of energy. If

$$G_{\mu\nu} = 0 \qquad \text{(6 conditions)}$$

space-time is curved in the first degree. This is the state of the world in an empty region—not containing matter, light or electromagnetic fields, but in the neighbourhood of these forms of energy.

If

$$G = 0 \qquad \text{(1 condition)}$$

space-time is curved in the second degree. This is the state of the world in a region not containing matter or electrons (bound energy), but containing light or electromagnetic fields (free energy).

If

$$G \text{ is not zero}$$

space-time is fully curved. This is the state of the world in a region containing continuous matter.

According to current physical theory continuous matter does not exist, so that strictly speaking the last case never arises. Matter is built of electrons or other nuclei. The regions lying between the electrons are not fully curved, whilst the regions inside the electrons must be cut out of space-time altogether. We cannot imagine ourselves exploring the inside of an electron with moving particles, light-waves, or material clocks and measuring-rods; hence, without further definition, any geometry of the interior, or any statement about space and time in the interior, is meaningless. But in common life, and frequently in

[7]This is not a recognised nomenclature.

physics, we are not concerned with this *microscopic* structure of matter. We need to know, not the actual values of the g's at a point, but their average values through a region, small from the ordinary standpoint but large compared with the molecular structure of matter. In this *macroscopic* treatment molecular matter is replaced by continuous matter, and uncurved space-time studded with holes is replaced by an equivalent fully curved space-time without holes.

It is natural that our senses should have developed faculties for perceiving some of these intrinsic distinctions of the possible states of the world around us. I prefer to think of matter and energy, not as agents causing the degrees of curvature of the world, but as parts of our perceptions of the existence of the curvature.

It will be seen that the law of gravitation can be summed up in the statement that in an empty region space-time can be curved only in the first degree.

6 The New Law of Gravitation and the Old Law

I don't know what I may seem to the world, but, as to myself, I seem to have been only as a boy playing on the sea-shore, and diverting myself in now and then finding a smoother pebble or a prettier shell than ordinary, whilst the great ocean of truth lay all undiscovered before me.

Sir Isaac Newton.

Was there any reason to feel dissatisfied with Newton's law of gravitation?

Observationally it had been subjected to the most stringent tests, and had come to be regarded as the perfect model of an exact law of nature. The cases, where a possible failure could be alleged, were almost insignificant. There are certain unexplained irregularities in the moon's motion; but astronomers generally looked—and must still look—in other directions for the cause of these discrepancies. One failure only had led to a serious questioning of the law; this was the discordance of motion of the perihelion of Mercury. How small was this discrepancy may be judged from the fact that, to meet it, it was proposed to amend *square* of the distance to the 2.00000016 power of the distance. Further it seemed possible, though unlikely, that the matter causing the zodiacal light might be of sufficient mass to be responsible for this effect.

The most serious objection against the Newtonian law as an exact law was that it had become ambiguous. The law refers to the product of the masses of the two bodies; but the mass depends on the velocity—a fact unknown in Newton's day. Are we to take the variable mass, or the mass reduced to rest? Perhaps a learned judge, interpreting Newton's statement like a last will and testament, could give a decision; but that is scarcely the way to settle an important point in scientific theory.

Further *distance*, also referred to in the law, is something relative to an observer. Are we to take the observer travelling with the sun or with the other body concerned, or at rest in the aether or in some gravitational medium?

Finally is the force of gravitation propagated instantaneously, or with the velocity of light, or some other velocity? Until comparatively recently it was thought that conclusive proof had been given that the speed of gravitation must be far higher than that of light. The argument was something like this. If the Sun attracts Jupiter towards its present position S, and Jupiter attracts the Sun towards its present position J, the two forces are in the same line and balance. But if the Sun attracts Jupiter towards its previous position S', and Jupiter attracts the Sun towards its previous position J', when the force of attraction started out to cross the gulf, then the two forces give a couple. This couple will

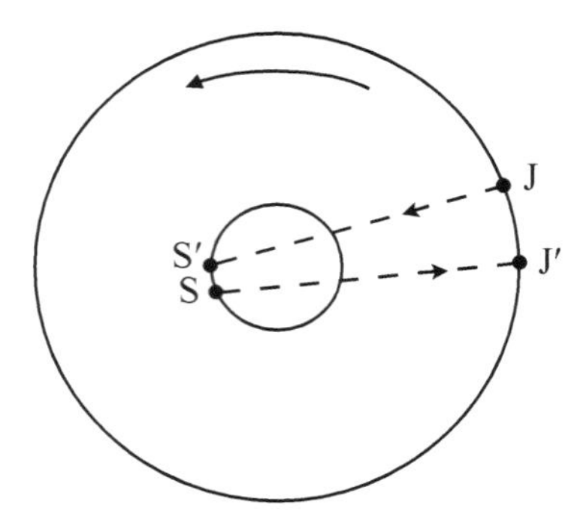

FIG. 13

tend to increase the angular momentum of the system, and, acting cumulatively, will soon cause an appreciable change of period, disagreeing with observation if the speed is at all comparable with that of light. The argument is fallacious, because the effect of propagation will not necessarily be that S is attracted in the direction towards J'. Indeed it is found that if S and J are two electric charges, S will be attracted very approximately towards J (not J') in spite of the electric influence being propagated with the velocity of light.[1]In the theory given in this book, gravitation is propagated with the speed of light, and there is no discordance with observation.

It is often urged that Newton's law of gravitation is much simpler than Einstein's new law. That depends on the point of view; and from the point of view of the four-dimensional world Newton's law is far more complicated. Moreover, it will be seen that if the ambiguities are to be cleared up, the statement of Newton's law must be greatly expanded.

Some attempts have been made to expand Newton's law on the basis of the restricted principle of relativity (p. 15) alone. This was insufficient to determine a definite amendment. Using the principle of equivalence, or relativity of force, we have arrived at a definite law proposed in the last chapter. Probably the question has arisen in the reader's mind, why should it be called the law of gravitation? It may be plausible as a law of nature; but what has the degree of curvature of space-time to do with attractive forces, whether real or apparent?

A race of flat-fish once lived in an ocean in which there were only two dimensions. It was noticed that in general fishes swam in straight lines, unless there was something obviously interfering with their free courses. This seemed a very natural behaviour. But there was a certain region where all the fish seemed to be bewitched; some passed through the region but changed the direction of their swim, others swam round and round indefinitely. One fish invented a theory of vortices, and said that there were whirlpools in that region which carried everything round in curves. By-and-by a far better theory was proposed; it was said that the fishes were all attracted towards a particularly large fish—a sun-fish—which was lying asleep in the middle of the region; and that was what caused

[1]Appendix, Note 6.

the deviation of their paths. The theory might not have sounded particularly plausible at first; but it was confirmed with marvellous exactitude by all kinds of experimental tests. All fish were found to possess this attractive power in proportion to their sizes; the law of attraction was extremely simple, and yet it was found to explain all the motions with an accuracy never approached before in any scientific investigations. Some fish grumbled that they did not see how there could be such an influence at a distance; but it was generally agreed that the influence was communicated through the ocean and might be better understood when more was known about the nature of water. Accordingly, nearly every fish who wanted to explain the attraction started by proposing some kind of mechanism for transmitting it through the water.

But there was one fish who thought of quite another plan. He was impressed by the fact that whether the fish were big or little they always took the same course, although it would naturally take a bigger force to deflect the bigger fish. He therefore concentrated attention on the courses rather than on the forces. And then he arrived at a striking explanation of the whole thing. There was a mound in the world round about where the sun-fish lay. Flat-fish could not appreciate it directly because they were two-dimensional; but whenever a fish went swimming over the slopes of the mound, although he did his best to swim straight on, he got turned round a bit. (If a traveller goes over the left slope of a mountain, he must consciously keep bearing away to the left if he wishes to keep to his original direction relative to the points of the compass.) This was the secret of the mysterious attraction, or bending of the paths, which was experienced in the region.

The parable is not perfect, because it refers to a hummock in space alone, whereas we have to deal with hummocks in space-time. But it illustrates how a curvature of the world we live in may give an illusion of attractive force, and indeed can only be discovered through some such effect. How this works out in detail must now be considered.

In the form $G_{\mu\nu} = 0$, Einstein's law expresses conditions to be satisfied in a gravitational field produced by any arbitrary distribution of attracting matter. An analogous form of Newton's law was given by Laplace in his celebrated expression $\nabla^2 V = 0$. A more illuminating form of the law is obtained if, instead of putting the question what kinds of space-time can exist under the most general conditions in an empty region, we ask what kind of space-time exists in the region round a single attracting particle? We separate out the effect of a single particle, just as Newton did. We can further simplify matters by introducing some definite mesh-system, which, of course, must be of a type which is not inconsistent with the kind of space-time found.

We need only consider space of two dimensions—sufficient for the so-called plane orbit of a planet—time being added as the third dimension. The remaining dimension of space can always be added, if desired, by conditions of symmetry. The result of long algebraic calculations[2] is that, round a particle

$$ds^2 = -\frac{1}{\gamma}\,dr^2 - r^2\,d\theta^2 + \gamma\,dt^2 \tag{6}$$

[2]Appendix, Note 7.

where $\gamma = 1 - \dfrac{2m}{r}$.

The quantity m is the gravitational mass of the particle—but we are not supposed to know that at present. r and θ are polar coordinates, the mesh-system being as in Fig. 11; or rather they are the nearest thing to polar coordinates that can be found in space which is not truly flat.

The fact is that this expression for ds^2 is found in the first place simply as a particular solution of Einstein's equations of the gravitational field; it is a variety of hummock (apparently the simplest variety) which is not curved beyond the first degree. There *could* be such a state of the world under suitable circumstances. To find out what those circumstances are, we have to trace some of the consequences, find out how any particle moves when ds^2 is of this form, and then examine whether we know of any case in which these consequences are found observationally. It is only after having ascertained that this form of ds^2 does correspond to the leading observed effects attributable to a particle of mass m at the origin that we have the right to identify this particular solution with the one we hoped to find.

It will be a sufficient illustration of this procedure, if we indicate how the position of the matter causing this particular solution is located. Wherever the formula (6) holds good there can be no matter, because the law which applies to empty space is satisfied. But if we try to approach the origin ($r = 0$), a curious thing happens. Suppose we take a measuring-rod, and, laying it radially, start marking off equal lengths with it along a radius, gradually approaching the origin. Keeping the time t constant, and $d\theta$ being zero for radial measurements, the formula (6) reduces to

$$ds^2 = -\frac{1}{\gamma}\, dr^2$$

or

$$dr^2 = -\gamma\, ds^2.$$

We start with r large. By-and-by we approach the point where $r = 2m$. But here, from its definition, γ is equal to 0. So that, however large the measured interval ds may be, $dr = 0$. We can go on shifting the measuring-rod through its own length time after time, but dr is zero; that is to say, we do not reduce r. There is a magic circle which no measurement can bring us inside. It is not unnatural that we should picture something obstructing our closer approach, and say that a particle of matter is filling up the interior.

The fact is that so long as we keep to space-time curved only in the first degree, we can never round off the summit of the hummock. It must end in an infinite chimney. In place of the chimney, however, we round it off with a small region of greater curvature. This region cannot be empty because the law applying to empty space does not hold. We describe it therefore as containing matter—a procedure which practically amounts to a definition of matter. Those familiar with hydrodynamics may be reminded of the problem of the irrotational rotation of a fluid; the conditions cannot be satisfied at the origin, and it is necessary to cut out a region which is filled by a vortex-filament.

A word must also be said as to the coordinates r and t used in (6). They correspond to our ordinary notion of radial distance and time—as well as any variables in a non-Euclidean world can correspond to words which, as ordinarily used, presuppose a Euclidean world. We shall thus call r and t, distance and time. But to give names to coordinates does not give more information—and in this case gives considerably less information—than is already contained in the formula for ds^2. If any question arises as to the exact significance of r and t it must always be settled by reference to equation (6).

The want of flatness in the gravitational field is indicated by the deviation of the coefficient γ from unity. If the mass $m = 0$, $\gamma = 1$, and space-time is perfectly flat. Even in the most intense gravitational fields known, the deviation is extremely small. For the sun, the quantity m, called the gravitational mass, is only 1.47 kilometres[3], for the earth it is 5 millimetres. In any practical problem the ratio $2m/r$ must be exceedingly small. Yet it is on the small corresponding difference in γ that the whole of the phenomena of gravitation depend.

The coefficient γ appears twice in the formula, and so modifies the flatness of space-time in two ways. But as a rule these two ways are by no means equally important. Its appearance as a coefficient of dt^2 produces much the most striking effects. Suppose that it is wished to measure the interval between two events in the history of a planet. If the events are, say 1 second apart in time, $dt = 1$ second $= 300,000$ kilometres. Thus $dt^2 = 90,000,000,000$ sq. km. Now no planet moves more than 50 kilometres in a second, so that the change dr associated with the lapse of 1 second in the history of the planet will not be more than 50 km. Thus dr^2 is not more than 2500 sq. km. Evidently the small term $2m/r$ has a much greater chance of making an impression where it is multiplied by dt^2 than where it is multiplied by dr^2.

Accordingly as a first approximation, we ignore the coefficient of dr^2, and consider only the meaning of

$$ds^2 = -dr^2 - r^2\,d\theta^2 + (1 - 2m/r)\,dt^2. \tag{7}$$

We shall now show that particles situated in this kind of space-time will appear to be under the influence of an attractive force directed towards the origin.

Let us consider the problem of mapping a small portion of this kind of world on a plane.

It is first necessary to define carefully the distinction which is here drawn between a "picture" and a "map." If we are given the latitudes and longitudes of a number of places on the earth, we can make a picture by taking latitude and longitude as vertical and horizontal distances, so that the lines of latitude and longitude form a mesh-system of squares; but that does not give a true map. In an ordinary map of Europe the lines of longitude run obliquely and the lines of latitude are curved. Why is this? Because the map aims at showing as accurately as possible all distances in their true proportions.[4] Distance is the important thing which it is desired to represent correctly. In four dimensions interval is the analogue of distance, and a map of the four-dimensional world will aim at showing all the intervals in their correct proportions. Our natural *picture* of space-time takes r and t as horizontal and vertical distances, e.g. when

[3]Appendix, Note 8.

[4]This is usually the object, though maps are sometimes made for a different purpose, e.g. Mercator's Chart.

we plot the graph of the motion of a particle; but in a true *map*, representing the intervals in their proper proportions, the r and t lines run obliquely or in curves across the map.

The instructions for drawing latitude and longitude lines (β, λ) on a map, are summed up in the formula for ds, p. 58,

$$ds^2 = d\beta^2 + \cos^2 \beta \, d\lambda^2,$$

and similarly the instructions for drawing the r and t lines are given by the

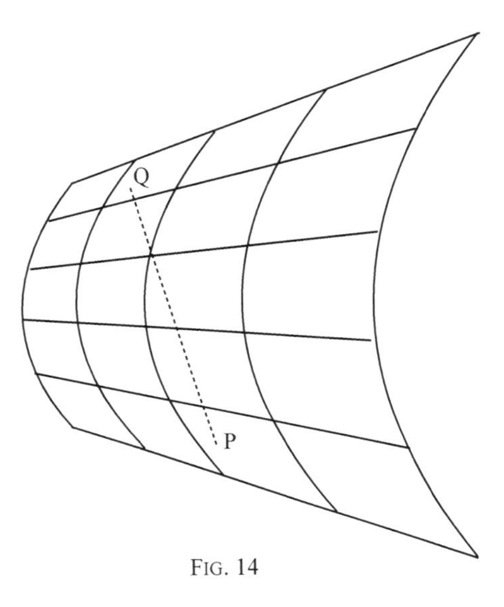

FIG. 14

formula (7).

The map is shown in Fig. 14. It is not difficult to see why the t-lines converge to the left of the diagram. The factor $1 - 2m/r$ decreases towards the left where r is small; and consequently any change of t corresponds to a shorter interval, and must be represented in the map by a shorter distance on the left. It is less easy to see why the r-lines take the courses shown; by analogy with latitude and longitude we might expect them to be curved the other way. But we discussed in Chapter 3 how the slope of the time-direction is connected with the slope of the space-direction; and it will be seen that the map gives approximately diamond-shaped partitions of the kind represented in Fig. 6.[5]

Like all maps of curved surfaces, the diagram is only accurate in the limit when the area covered is very small.

It is important to understand clearly the meaning of this map. When we speak in the ordinary way of distance from the sun and the time at a point in the solar system, we mean the two variables r and t. These are not the result of any precise measures with scales and clocks made at a point, but are mathematical variables most appropriate for describing the whole solar system. They represent a compromise, because it is necessary to deal with a region too

[5] The substitution $x = r + \frac{1}{2}t^2 m/r^2$, $y = t(1 - m/r)$, gives $ds^2 = -dx^2 + dy^2$, if squares of m are negligible. The map is drawn with x and y as rectangular coordinates.

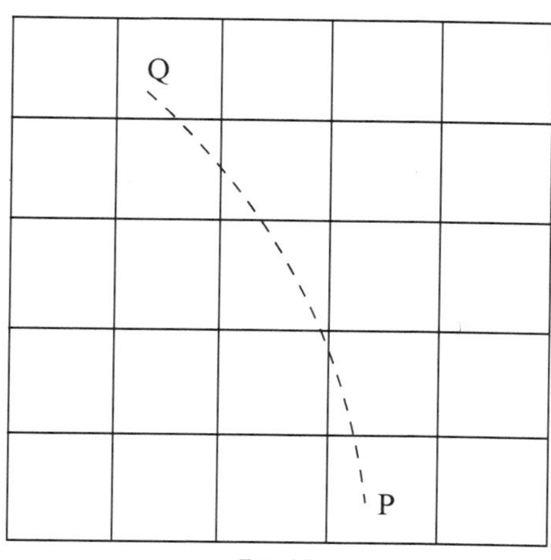

FIG. 15

large for accurate representation on a plane map. We should naturally picture them as rectangular coordinates partitioning space-time into square meshes, as in Fig. 15; but such a picture is not a true map, because it does not represent in their true proportions the intervals between the various points in the picture. It is not possible to draw any map of the whole curved region without distortion; but a small enough portion can be represented without distortion if the partitions of equal r and t are drawn as in Fig. 14. To get back from the true map to the customary picture of r and t as perpendicular space and time, we must strain Fig. 14 until all the meshes become squares as in Fig. 15.

Now in the map the geometry is Euclidean and the tracks of all material particles will be straight lines. Take such a straight track PQ, which will necessarily be nearly vertical, unless the velocity is very large. Strain the figure so as to obtain the customary representation of r and t (in Fig. 15), and the track PQ will become curved—curved towards the left, where the sun lies. In each successive vertical interval (time), a successively greater progress is made to the left horizontally (space). Thus the velocity towards the sun increases. We say that the particle is attracted to the sun.

The mathematical reader should find no difficulty in proving from the diagram that for a particle with small velocity the acceleration towards the sun is approximately m/r^2, agreeing with the Newtonian law.

Tracks for very high speeds may be affected rather differently. The track corresponding to a wave of light is represented by a straight line at 45 to the horizontal in Fig. 14. It would require very careful drawing to trace what happens to it when the strain is made transforming to Fig. 15; but actually, whilst becoming more nearly vertical, it receives a curvature in the opposite direction. The effect of the gravitation of the sun on a light-wave, or very fast particle, proceeding radially is actually a *repulsion*!

The track of a transverse light-wave, coming out from the plane of the paper, will be affected like that of a particle of zero velocity in distorting from Fig. 14 to Fig. 15. Hence the sun's influence on a transverse light-wave is always an attraction. The acceleration is simply m/r^2 as for a particle at rest.

The result that the expression found for the geometry of the gravitational field of a particle leads to Newton's law of attraction is of great importance. It shows that the law, $G_{\mu\nu} = 0$, proposed on theoretical grounds, agrees with observation at least approximately. It is no drawback that the Newtonian law applies only when the speed is small; all planetary speeds are small compared with the velocity of light, and the considerations mentioned at the beginning of this chapter suggest that some modification may be needed for speeds comparable with that of light.

Another important point to notice is that the attraction of gravitation is simply a geometrical deformation of the straight tracks. It makes no difference what body or influence is pursuing the track, the deformation is a general discrepancy between the "mental picture" and the "true map" of the portion of space-time considered. Hence light is subject to the same disturbance of path as matter. This is involved in the Principle of Equivalence; otherwise we could distinguish between the acceleration of a lift and a true increase of gravitation by optical experiments; in that case the observer for whom light-rays appear to take straight tracks might be described as absolutely unaccelerated and there could be no relativity theory. Physicists in general have been prepared to admit the likelihood of an influence of gravitation on light similar to that exerted on matter; and the problem whether or not light has "weight" has often been considered.

The appearance of γ as the coefficient of dt^2 is responsible for the main features of Newtonian gravitation; the appearance of $1/\gamma$ as the coefficient of dr^2 is responsible for the principal deviations of the new law from the old. This classification seems to be correct; but the Newtonian law is ambiguous and it is difficult to say exactly what are to be regarded as discrepancies from it. Leaving aside now the time-term as sufficiently discussed, we consider the space-terms alone[6]

$$ds^2 = \frac{1}{\gamma}\, dr^2 + r^2\, d\theta^2.$$

The expression shows that space considered alone is non-Euclidean in the neighbourhood of an attracting particle. This is something entirely outside the scope of the old law of gravitation. Time can only be explored by something moving, whether a free particle or the parts of a clock, so that the non-Euclidean character of space-time can be covered up by introducing a field of force, suitably modifying the motion, as a convenient fiction. But space can be explored by static methods; and theoretically its non-Euclidean character could be ascertained by sufficiently precise measures with rigid scales.

If we lay our measuring scale transversely and proceed to measure the circumference of a circle of nominal radius r, we see from the formula that the measured length ds is equal to $r\,d\theta$, so that, when we have gone right round the circle, θ has increased by 2π and the measured circumference is $2\pi r$. But when we lay the scale radially the measured length ds is equal to $dr/\sqrt{\gamma}$, which is always greater than dr. Thus, in measuring a diameter, we obtain a result greater than $2r$, each portion being greater than the corresponding change of r.

[6] We change the sign of ds^2, so that ds, when real, means measured space instead of measured time.

Thus if we draw a circle, placing a massive particle near the centre so as to produce a gravitational field, and measure with a rigid scale the circumference and the diameter, the ratio of the measured circumference to the measured diameter will not be the famous number $\pi = 3.14159265358979323846264338327 9\ldots$ but a little smaller. Or if we inscribe a regular hexagon in this circle its sides will not be exactly equal to the radius of the circle. Placing the particle near, instead of at, the centre, avoids measuring the diameter *through* the particle, and so makes the experiment a practical one. But though practical, it is not practicable to determine the non-Euclidean character of space in this way. Sufficient refinement of measures is not attainable. If the mass of a ton were placed inside a circle of 5 yards radius, the defect in the value of π would only appear in the twenty-fourth or twenty-fifth place of decimals.

It is of value to put the result in this way, because it shows that the relativist is not talking metaphysics when he says that space in the gravitational field is non-Euclidean. His statement has a plain physical meaning, which we may some day learn how to test experimentally. Meanwhile we can test it by indirect methods.

Suppose that a plane field is uniformly studded with hurdles. The distance between any two points will be proportional to the number of hurdles that must be passed over in getting from one point to the other by the straight route—in fact the minimum number of hurdles. We can use counts of hurdles as the equivalent of distance, and map the field by these counts. The map can be drawn on a plane sheet of paper without any inconsistency, since the field is plane. Let us now dismiss from our minds all idea of distances in the field or straight lines in the field, and assume that distances on the map merely represent the minimum number of hurdles between two points; straight lines on the map will represent the corresponding routes. This has the advantage that if an earthquake occurs, deforming the field, the map will still be correct. The path of fewest hurdles will still cross the same hurdles as before the earthquake; it will be twisted out of the straight line in the field; but we should gain nothing by taking a straighter course, since that would lead through a region where the hurdles are more crowded. We do not alter the number of hurdles in any path by deforming it.

This can be illustrated by Figs. 14 and 15. Fig. 14 represents the original undistorted field with the hurdles uniformly placed. The straight line PQ represents the path of fewest hurdles from P to Q, and its length is proportional to the number of hurdles. Fig. 15 represents the distorted field, with PQ distorted into a curve; but PQ is still the path of fewest hurdles from P to Q, and the number of hurdles in the path is the same as before. If therefore we map according to hurdle-counts we arrive at Fig. 14 again, just as though no deformation had taken place.

To make any difference in the hurdle-counts, the hurdles must be taken up and replanted. Starting from a given point as centre, let us arrange them so that they gradually thin out towards the boundaries of the field. Now choose a circle with this point as centre;—but first, what is a circle? It has to be defined in terms of hurdle-counts; and clearly it must be a curve such that the minimum number of hurdles between any point on it and the centre is a constant (the radius). With this definition we can defy earthquakes. The number of hurdles in the circumference of such a circle will not bear the same proportion to the number in the radius as in the field of uniform hurdles; owing to the crowding near the centre, the ratio will be less. Thus we have a suitable analogy for a circle whose circumference is less than π times its diameter.

This analogy enables us to picture the condition of space round a heavy particle, where the ratio of the circumference of a circle to the diameter is less than π. Hurdle-counts will no longer be accurately mappable on a plane sheet of paper, because they do not conform to Euclidean geometry.

Now suppose a heavy particle wishes to cross this field, passing near but not through the centre. In Euclidean space, with the hurdles uniformly distributed, it travels in a straight line, i.e. it goes between any two points by a path giving the fewest hurdle jumps. We may assume that in the non-Euclidean field with rearranged hurdles, the particle still goes by the path of least effort. In fact, in any small portion we cannot distinguish between the rearrangement and a distortion; so we may imagine that the particle takes each portion as it comes according to the rule, and is not troubled by the rearrangement which is only visible to a general survey of the whole field[7].

Now clearly it will pay not to go straight through the dense portion, but to keep a little to the outside where the hurdles are sparser—not too much, or the path will be unduly lengthened. The particle's track will thus be a little concave to the centre, and an onlooker will say that it has been attracted to the centre. It is rather curious that we should call it attraction, when the track has rather been avoiding the central region; but it is clear that the direction of motion has been bent round in the way attributable to an attractive force.

This bending of the path is additional to that due to the Newtonian force of gravitation which depends on the second appearance of γ in the formula. As already explained it is in general a far smaller effect and will appear only as a minute correction to Newton's law. The only case where the two rise to equal importance is when the track is that of a light-wave, or of a particle moving with a speed approaching that of light; for then dr^2 rises to the same order of magnitude as dt^2.

To sum up, a ray of light passing near a heavy particle will be bent, firstly, owing to the non-Euclidean character of the combination of time with space. This bending is equivalent to that due to Newtonian gravitation, and may be calculated in the ordinary way on the assumption that light has weight like a material body. Secondly, it will be bent owing to the non-Euclidean character of space alone, and this curvature is additional to that predicted by Newton's law. If then we can observe the amount of curvature of a ray of light, we can make a crucial test of whether Einstein's or Newton's theory is obeyed.

This separation of the attraction into two parts is useful in a comparison of the new theory with the old; but from the point of view of relativity it is artificial. Our view is that light is bent just in the same way as the track of a material particle moving with the same velocity would be bent. Both causes of bending may be ascribed either to weight or to non-Euclidean space-time, according to

[7] There must be some absolute track, and if absolute significance can only be associated with hurdle-counts and not with distances in the field, the path of fewest hurdles is the only track capable of absolute definition.

the nomenclature preferred. The only difference between the predictions of the old and new theories is that in one case the weight is calculated according to Newton's law of gravitation, in the other case according to Einstein's.

There is an alternative way of viewing this effect on light according to Einstein's theory, which, for many reasons is to be preferred. This depends on the fact that the velocity of light in the gravitational field is not a constant (unity) but becomes smaller as we approach the sun. This does not mean that an observer determining the velocity of light experimentally at a spot near the sun would detect the decrease; if he performed Fizeau's experiment, his result in kilometres per second would be exactly the same as that of a terrestrial observer. It is the coordinate velocity that is here referred to, described in terms of the quantities r, θ, t, introduced by the observer who is contemplating the whole solar system at the same time.

It will be remembered that in discussing the approximate geometry of space-time in Fig. 3, we found that certain events like P were in the absolute past or future of O, and others like P' were neither before nor after O, but elsewhere. Analytically the distinction is that for the interval OP, ds^2 is positive; for OP', ds^2 is negative. In the first case the interval is real or "time-like"; in the second it is imaginary or "space-like." The two regions are separated by lines (or strictly, cones) in crossing which ds^2 changes from positive to negative; and along the lines themselves ds is zero. It is clear that these lines must have important absolute significance in the geometry of the world. Physically their most important property is that pulses of light travel along these tracks, and the motion of a light-pulse is always given by the equation $ds = 0$.

Using the expression for ds^2 in a gravitational field, we accordingly have for light

$$0 = -\frac{1}{\gamma}\, dr^2 - r^2\, d\theta^2 + \gamma\, dt^2.$$

For radial motion, $d\theta = 0$, and therefore

$$\left(\frac{dr}{dt}\right)^2 = \gamma^2.$$

For transverse motion, $dr = 0$, and therefore

$$\left(\frac{r\, d\theta}{dt}\right)^2 = \gamma.$$

Thus the coordinate velocity of light travelling radially is γ, and of light travelling transversely is $\surd\gamma$, in the coordinates chosen.

The coordinate velocity must depend on the coordinates chosen; and it is more convenient to use a slightly different system in which the velocity of light is the same in all directions[8], viz. γ or $1 - 2m/r$. This diminishes as we approach

[8]This is obtained by writing $r + m$ instead of r, or diminishing the nominal distance of the sun by $1\frac{1}{2}$ kilometres. This change of coordinates simplifies the problem, but can, of course, make no difference to anything observable. After we have traced the course of the light ray in the coordinates chosen, we have to connect the results with experimental measures, using the corresponding formula for ds^2. This final connection of mathematical and experimental results is, however, comparatively simple, because it relates to measuring operations performed in a terrestrial observatory where the difference of γ from unity is negligible.

the sun—an illustration of our previous remark that a pulse of light proceeding radially is repelled by the sun.

The wave-motion in a ray of light can be compared to a succession of long straight waves rolling onward in the sea. If the motion of the waves is slower at one end than the other, the whole wave-front must gradually slew round, and the direction in which it is rolling must change. In the sea this happens when one end of the wave reaches shallow water before the other, because the speed in shallow water is slower. It is well known that this causes waves proceeding diagonally across a bay to slew round and come in parallel to the shore; the advanced end is delayed in the shallow water and waits for the other. In the same way when the light waves pass near the sun, the end nearest the sun has the smaller velocity and the wave-front slews round; thus the course of the waves is bent.

Light moves more slowly in a material medium than in vacuum, the velocity being inversely proportional to the refractive index of the medium. The phenomenon of refraction is in fact caused by a slewing of the wave-front in passing into a region of smaller velocity. We can thus imitate the gravitational effect on light precisely, if we imagine the space round the sun filled with a refracting medium which gives the appropriate velocity of light. To give the velocity $1 - 2m/r$, the refractive index must be $1/(1 - 2m/r)$, or, very approximately, $1 + 2m/r$. At the surface of the sun, $r = 697,000$ km., $m = 1.47$ km., hence the necessary refractive index is 1.00000424. At a height above the sun equal to the radius it is 1.00000212.

Any problem on the paths of rays near the sun can now be solved by the methods of geometrical optics applied to the equivalent refracting medium. It is not difficult to show that the total deflection of a ray of light passing at a distance r from the centre of the sun is (in circular measure)

$$\frac{4m}{r},$$

whereas the deflection of the same ray calculated on the Newtonian theory would be

$$\frac{2m}{r}.$$

For a ray grazing the surface of the sun the numerical value of this deflection is

$$\begin{aligned} &1''.75 \quad \text{(Einstein's theory)}, \\ &0''.87 \quad \text{(Newtonian theory)}. \end{aligned}$$

7 WEIGHING LIGHT

Query 1. Do not Bodies act upon Light at a distance, and by their action bend its Rays, and is not this action (*caeteris paribus*) strongest at the least distance?

NEWTON, *Opticks*.

WE come now to the experimental test of the influence of gravitation on light discussed theoretically in the last chapter. It is not the general purpose of this book to enter into details of experiments; and if we followed this plan consistently, we should, as hitherto, summarise the results of the observations in a few lines. But it is this particular test which has turned public attention towards the relativity theory, and there appears to be widespread desire for information. We shall therefore tell the story of the eclipse expeditions in some detail. It will make a break in the long theoretical arguments, and will illustrate the important applications of this theory to practical observations.

It must be understood that there were two questions to answer: firstly, whether light has weight (as suggested by Newton), or is indifferent to gravitation; secondly, if it has weight, is the amount of the deflection in accordance with Einstein's or Newton's laws?

It was already known that light possesses mass or inertia like other forms of electromagnetic energy. This is manifested in the phenomena of radiation-pressure. Some force is required to stop a beam of light by holding an obstacle in its path; a searchlight experiences a minute force of recoil just as if it were a machine-gun firing material projectiles. The force, which is predicted by orthodox electromagnetic theory, is exceedingly minute; but delicate experiments have detected it. Probably this inertia of radiation is of great cosmical importance, playing a great part in the equilibrium of the more diffuse stars. Indeed it is probably the agent which has carved the material of the universe into stars of roughly uniform mass. Possibly the tails of comets are a witness to the power of the momentum of sunlight, which drives outwards the smaller or the more absorptive particles.

It is legitimate to speak of a pound of light as we speak of a pound of any other substance. The mass of ordinary quantities of light is however extremely small, and I have calculated that at the low charge of 3*d.* a unit, an Electric Light Company would have to sell light at the rate of £140,000,000 a pound. All the sunlight falling on the earth amounts to 160 tons daily.

It is perhaps not easy to realise how a wave-motion can have inertia, and it is still more difficult to understand what is meant by its having weight. Perhaps this will be better understood if we put the problem in a concrete form. Imagine a hollow body, with radiant heat or light-waves traversing the hollow; the mass of the body will be the sum of the masses of the material and of the radiant energy in the hollow; a greater force will be required to shift it because of the light-waves contained in it. Now let us weigh it with scales or a spring-balance. Will it also weigh heavier on account of the radiation contained, or will the weight be that of the solid material alone? If the former, then clearly from this aspect light has weight; and it is not difficult to deduce the effect of this weight on a freely moving light-beam not enclosed within a hollow.

The effect of weight is that the radiation in the hollow body acquires each second a downward momentum proportional to its mass. This in the long run is transmitted to the material enclosing it. For a free light-wave in space, the added momentum combines with the original momentum, and the total momentum determines the direction of the ray, which is accordingly bent. Newton's theory suggests no means for bringing about the bending, but contents itself with predicting it on general principles. Einstein's theory provides a means, viz. the variation of velocity of the waves.

Hitherto mass and weight have always been found associated in strict proportionality. One very important test had already shown that this proportionality is not confined to material energy. The substance uranium contains a great deal of radio-active energy, presumably of an electromagnetic nature, which it slowly liberates. The mass of this energy must be an appreciable fraction of the whole mass of the substance. But it was shown by experiments with the Eötvös torsion-balance that the ratio of weight to mass for uranium is the same as for all other substances; so the energy of radio-activity has weight. Still even this experiment deals only with bound electromagnetic energy, and we are not justified in deducing the properties of the free energy of light.

It is easy to see that a terrestrial experiment has at present no chance of success. If the mass and weight of light are in the same proportion as for matter, the ray of light will be bent just like the trajectory of a material particle. On the earth a rifle bullet, like everything else, drops 16 feet in the first second, 64 feet in two seconds, and so on, below its original line of flight; the rifle must thus be aimed above the target. Light would also drop 16 feet in the first second[1]; but, since it has travelled 186,000 miles along its course in that time, the bend is inappreciable. In fact any terrestrial course is described so quickly

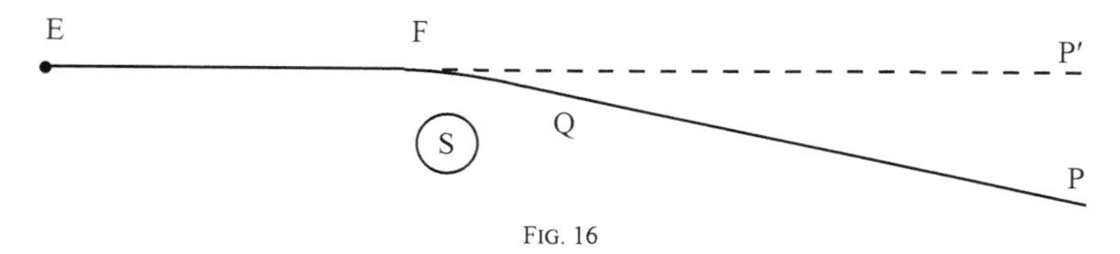

FIG. 16

that gravitation has scarcely had time to accomplish anything.

The experiment is therefore transferred to the neighbourhood of the sun. There we get a pull of gravitation 27 times more intense than on the earth;

[1]Or 32 feet according to Einstein's law. The fall increases with the speed of the motion.

and—what is more important—the greater size of the sun permits a much longer trajectory throughout which the gravitation is reasonably powerful. The deflection in this case may amount to something of the order of a second of arc, which for the astronomer is a fairly large quantity.

In Fig. 16 the line $EFQP$ shows the track of a ray of light from a distant star P which reaches the earth E. The main part of the bending of the ray occurs as it passes the sun S; and the initial course PQ and the final course FE are practically straight. Since the light rays enter the observer's eye or telescope in the direction FE, this will be the direction in which the star appears. But its true direction from the earth is QP, the initial course. So the star appears displaced outwards from its true position by an angle equal to the total deflection of the light.

It must be noticed that this is only true because a star is so remote that its true direction with respect to the earth E is indistinguishable from its direction with respect to the point Q. For a source of light within the solar system, the apparent displacement of the source is by no means equal to the deflection of the light-ray. It is perhaps curious that the attraction of light by the sun should produce an apparent displacement of the star away from the sun; but the necessity for this is clear.

The bending affects stars seen near the sun, and accordingly the only chance of making the observation is during a total eclipse when the moon cuts off the dazzling light. Even then there is a great deal of light from the sun's corona which stretches far above the disc. It is thus necessary to have rather bright stars near the sun, which will not be lost in the glare of the corona. Further the displacements of these stars can only be measured relatively to other stars, preferably more distant from the sun and less displaced; we need therefore a reasonable number of outer bright stars to serve as reference points.

In a superstitious age a natural philosopher wishing to perform an important experiment would consult an astrologer to ascertain an auspicious moment for the trial. With better reason, an astronomer to-day consulting the stars would announce that the most favourable day of the year for weighing light is May 29. The reason is that the sun in its annual journey round the ecliptic goes through fields of stars of varying richness, but on May 29 it is in the midst of a quite exceptional patch of bright stars—part of the Hyades—by far the best star-field encountered. Now if this problem had been put forward at some other period of history, it might have been necessary to wait some thousands of years for a total eclipse of the sun to happen on the lucky date. But by strange good fortune an eclipse did happen on May 29, 1919. Owing to the curious sequence of eclipses a similar opportunity will recur in 1938; we are in the midst of the most favourable cycle. It is not suggested that it is impossible to make the test at other eclipses; but the work will necessarily be more difficult.

Attention was called to this remarkable opportunity by the Astronomer Royal in March, 1917; and preparations were begun by a Committee of the Royal Society and Royal Astronomical Society for making the observations. Two expeditions were sent to different places on the line of totality to minimise the risk of failure by bad weather. Dr A. C. D. Crommelin and Mr C. Davidson went to Sobral in North Brazil; Mr E. T. Cottingham and the writer went to the Isle of Principe in the Gulf of Guinea, West Africa. The instrumental equipment for both expeditions was prepared at Greenwich Observatory under the care of the Astronomer Royal; and here Mr Davidson made the arrangements which were the main factor in the success of both parties.

The circumstances of the two expeditions were somewhat different and it is scarcely possible to treat them together. We shall at first follow the fortunes of the Principe observers. They had a telescope of focal length 11 feet 4 inches. On their photographs 1 second of arc (which was about the largest displacement to be measured) corresponds to about $\frac{1}{1500}$ inch—by no means an inappreciable quantity. The aperture of the object-glass was 13 inches, but as used it was stopped down to 8 inches to give sharper images. It is necessary, even when the exposure is only a few seconds, to allow for the diurnal motion of the stars across the sky, making the telescope move so as to follow them. But since it is difficult to mount a long and heavy telescope in the necessary manner in a temporary installation in a remote part of the globe, the usual practice at eclipses is to keep the telescope rigid and reflect the stars into it by a coelostat—a plane mirror kept revolving at the right rate by clock-work. This arrangement was adopted by both expeditions.

The observers had rather more than a month on the island to make their preparations. On the day of the eclipse the weather was unfavourable. When totality began the dark disc of the moon surrounded by the corona was visible through cloud, much as the moon often appears through cloud on a night when no stars can be seen. There was nothing for it but to carry out the arranged programme and hope for the best. One observer was kept occupied changing the plates in rapid succession, whilst the other gave the exposures of the required length with a screen held in front of the object-glass to avoid shaking the telescope in any way.

> For in and out, above, about, below
> 'Tis nothing but a Magic *Shadow*-show
> Played in a Box whose candle is the Sun
> Round which we Phantom Figures come and go.

Our shadow-box takes up all our attention. There is a marvellous spectacle above, and, as the photographs afterwards revealed, a wonderful prominence-flame is poised a hundred thousand miles above the surface of the sun. We have no time to snatch a glance at it. We are conscious only of the weird half-light of the landscape and the hush of nature, broken by the calls of the observers, and beat of the metronome ticking out the 302 seconds of totality.

Sixteen photographs were obtained, with exposures ranging from 2 to 20 seconds. The earlier photographs showed no stars, though they portrayed the remarkable prominence; but apparently the cloud lightened somewhat towards the end of totality, and a few images appeared on the later plates. In many cases one or other of the most essential stars was missing through cloud, and no use could be made of them; but one plate was found showing fairly good images of five stars, which were suitable for a determination. This was measured on the spot a few days after the eclipse in a micrometric measuring-machine. The problem was to determine how the apparent positions of the stars, affected by the sun's gravitational field, compared with the normal positions on a photograph taken when the sun was out of the way. Normal photographs for comparison had been taken with the same telescope in England in January. The eclipse photograph and a comparison photograph were placed film to film in the measuring-machine so that corresponding images fell close together[2], and the small distances were measured in two rectangular directions. From these the relative displacements of the stars could be ascertained. In comparing two plates, various allowances have to be made for refraction, aberration, plate-orientation, etc.; but since these occur equally in determinations of stellar parallax, for which much greater accuracy is required, the necessary procedure is well-known to astronomers.

The results from this plate gave a definite displacement, in good accordance with Einstein's theory and disagreeing with the Newtonian prediction. Although the material was very meagre compared with what had been hoped for, the writer (who it must be admitted was not altogether unbiassed) believed it convincing.

It was not until after the return to England that any further confirmation was forthcoming. Four plates were brought home undeveloped, as they were of a brand which would not stand development in the hot climate. One of these was found to show sufficient stars; and on measurement it also showed the deflection predicted by Einstein, confirming the other plate.

The bugbear of possible systematic error affects all investigations of this kind. How do you know that there is not something in your apparatus responsible for this apparent deflection? Your object-glass has been shaken up by travelling; you have introduced a mirror into your optical system; perhaps the 50 rise of temperature between the climate at the equator and England in winter has done some kind of mischief. To meet this criticism, a different field of stars was photographed at night in Principe and also in England at the same altitude as the eclipse field. If the deflection were really instrumental, stars on these plates should show relative displacements of a similar kind to those on the eclipse plates. But on measuring these check-plates no appreciable displacements were found. That seems to be satisfactory evidence that the displacement observed during the eclipse is really due to the sun being in the region, and is not due to differences in instrumental conditions between England and Principe. Indeed the only possible loophole is a difference between the night conditions at Principe when the check-plates were taken, and the day, or rather eclipse, conditions when the eclipse photographs were taken. That seems impossible since the temperature at Principe did not vary more than 1 between day and night.

The problem appeared to be settled almost beyond doubt; and it was with some confidence that we awaited the return of the other expedition from Brazil. The Brazil party had had fine weather and had gained far more extensive material on their plates. They had remained two months after the eclipse to photograph the same region before dawn, when clear of the sun, in order that they might have comparison photographs taken under exactly the same circumstances. One set of photographs was secured with a telescope similar to that used at Principe. In addition they used a longer telescope of 4 inches aperture and 19 feet focal length[3].

[2]This was possible because at Principe the field of stars was reflected in the coelostat mirror, whereas in England it was photographed direct.

The photographs obtained with the former were disappointing. Although the full number of stars expected (about 12) were shown, and numerous plates had been obtained, the definition of the images had been spoiled by some cause, probably distortion of the coelostat-mirror by the heat of the sunshine falling on it. The observers were pessimistic as to the value of these photographs; but they were the first to be measured on return to England, and the results came as a great surprise after the indications of the Principe plates. The measures pointed with all too good agreement to the "half-deflection," that is to say, the Newtonian value which is one-half the amount required by Einstein's theory. It seemed difficult to pit the meagre material of Principe against the wealth of data secured from the clear sky of Sobral. It is true the Sobral images were condemned, but whether so far as to invalidate their testimony on this point was not at first clear; besides the Principe images were not particularly well-defined, and were much enfeebled by cloud. Certain compensating advantages of the latter were better appreciated later. Their strong point was the satisfactory check against systematic error afforded by the photographs of the check-field; there were no check-plates taken at Sobral, and, since it was obvious that the discordance of the two results depended on systematic error and not on the wealth of material, this distinctly favoured the Principe results. Further, at Principe there could be no evil effects from the sun's rays on the mirror, for the sun had withdrawn all too shyly behind the veil of cloud. A further advantage was provided by the check-plates at Principe, which gave an independent determination of the difference of scale of the telescope as used in England and at the eclipse; for the Sobral plates this scale-difference was eliminated by the method of reduction, with the consequence that the results depended on the measurement of a much smaller relative displacement.

There remained a set of seven plates taken at Sobral with the 4-inch lens; their measurement had been delayed by the necessity of modifying a micrometer to hold them, since they were of unusual size. From the first no one entertained any doubt that the final decision must rest with them, since the images were almost ideal, and they were on a larger scale than the other photographs. The use of this instrument must have presented considerable difficulties—the unwieldy length of the telescope, the slower speed of the lens necessitating longer exposures and more accurate driving of the clock-work, the larger scale rendering the focus more sensitive to disturbances—but the observers achieved success, and the perfection of the negatives surpassed anything that could have been hoped for.

These plates were now measured and they gave a final verdict definitely confirming Einstein's value of the deflection, in agreement with the results obtained at Principe.

It will be remembered that Einstein's theory predicts a deflection of $1''.74$ at the edge of the sun[4], the amount falling off inversely as the distance from the sun's centre. The simple Newtonian deflection is half this, $0''.87$. The final results (reduced to the edge of the sun) obtained at Sobral and Principe with

[3]See Frontispiece. The two telescopes are shown and the backs of the two coelostat-mirrors which reflect the sky into them. The clock driving the larger mirror is seen on the pedestal on the left.

[4]The predicted deflection of light from infinity to infinity is just over $1''.745$, from infinity to the earth it is just under.

their "probable accidental errors" were

Sobral $1''.980''.12$,
Principe $1''.610''.30$.

It is usual to allow a margin of safety of about twice the probable error on either side of the mean. The evidence of the Principe plates is thus just about sufficient to rule out the possibility of the "half-deflection," and the Sobral plates exclude it with practical certainty. The value of the material found at Principe cannot be put higher than about one-sixth of that at Sobral; but it certainly makes it less easy to bring criticism against this confirmation of Einstein's theory seeing that it was obtained independently with two different instruments at different places and with different kinds of checks.

The best check on the results obtained with the 4-inch lens at Sobral is the striking internal accordance of the measures for different stars. The theoretical deflection should vary inversely as the distance from the sun's centre; hence, if we plot the mean radial displacement found for each star separately against the inverse distance, the points should lie on a straight line. This is shown in Fig. 17

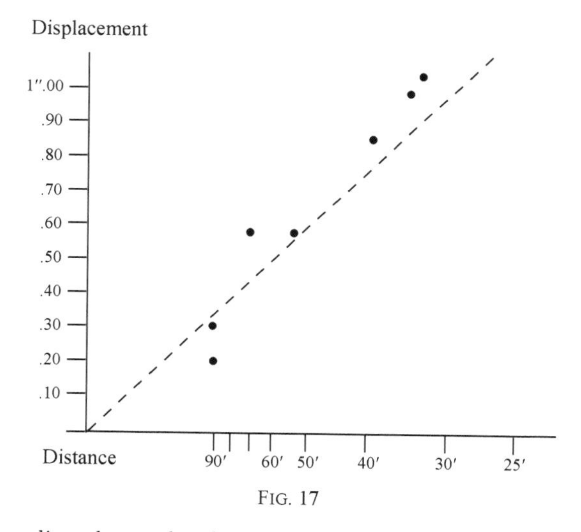

FIG. 17

where the broken line shows the theoretical prediction of Einstein, the deviations being within the accidental errors of the determinations. A line of half the slope representing the half-deflection would clearly be inadmissible.

Moreover, values of the deflection were deduced from the measures in right ascension and declination independently. These were in close agreement.

A diagram showing the relative positions of the stars is given in Fig. 18.

The square shows the limits of the plates used at Principe, and the oblique rectangle the limits with the 4-inch lens at Sobral. The centre of the sun moved from S to P in the $2\frac{1}{4}$ hours interval between totality at the two stations; the sun is here represented for a time about midway between. The stars measured on the Principe plates were Nos. 3, 4, 5, 6, 10, 11; those at Sobral were 11, 10, 6, 5, 4, 2, 3 (in the order of the dots from left to right in Fig. 17). None of these were fainter than $6^{m}{\cdot}0$, the brightest κ^1 Tauri (No. 4) being $4^{m}{\cdot}5$.

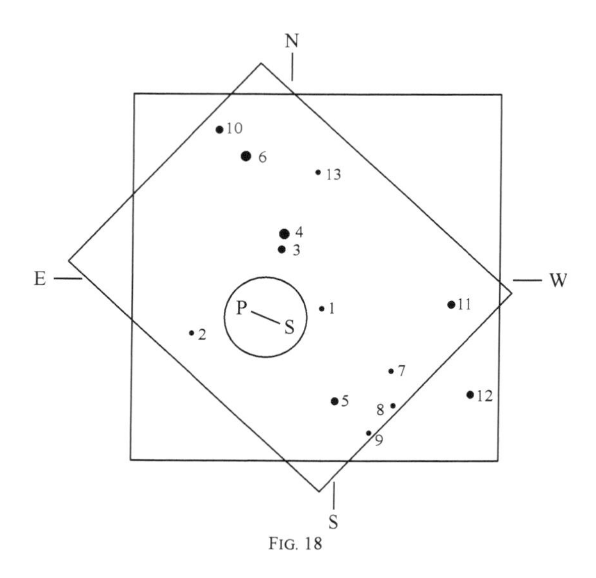

FIG. 18

It has been objected that although the observations establish a deflection of light in passing the sun equal to that predicted by Einstein, it is not immediately obvious that this deflection must necessarily be attributed to the sun's gravitational field. It is suggested that it may not be an essential effect of the sun as a massive body, but an accidental effect owing to the circumstance that the sun is surrounded by a corona which acts as a refracting atmosphere. It would be a strange coincidence if this atmosphere imitated the theoretical law in the exact quantitative way shown in Fig. 17; and the suggestion appears to us far-fetched. However the objection can be met in a more direct way. We have already shown that the gravitational effect on light is equivalent to that produced by a refracting medium round the sun and have calculated the necessary refractive index. At a height of 400,000 miles above the surface the refractive index required is 1.0000021. This corresponds to air at $\frac{1}{140}$ atmosphere, hydrogen at $\frac{1}{70}$ atmosphere, helium at $\frac{1}{20}$ atmospheric pressure. It seems obvious that there can be no material of this order of density at such a distance from the sun. The pressure on the sun's surface of the columns of material involved would be of the order 10,000 atmospheres; and we know from spectroscopic evidence that there is no pressure of this order. If it is urged that the mass could perhaps be supported by electrical forces, the argument from absorption is even more cogent. The light from the stars photographed during the eclipse has passed through a depth of at least a million miles of material of this order of density—or say the equivalent of 10,000 miles of air at atmospheric density. We know to our cost what absorption the earth's 5 miles of homogeneous atmosphere can effect. And yet at the eclipse the stars appeared on the photographs with their normal brightness. If the irrepressible critic insists that the material round the sun may be composed of some new element with properties unlike any material known to us, we may reply that the mechanism of refraction and of absorption is the same, and there is a limit to the possibility of refraction without appreciable absorption. Finally it would be necessary to arrange that the density of the material falls off inversely as the distance from the sun's centre in order to give the required variation of refractive index.

Several comets have been known to approach the sun within the limits of distance here considered. If they had to pass through an atmosphere of the density required to account for the displacement, they would have suffered enormous resistance. Dr Crommelin has shown that a study of these comets sets an upper limit to the density of the corona, which makes the refractive effect quite negligible.

Those who regard Einstein's law of gravitation as a natural deduction from a theory based on the minimum of hypotheses will be satisfied to find that his remarkable prediction is quantitatively confirmed by observation, and that no unforeseen cause has appeared to invalidate the test.

8 OTHER TESTS OF THE THEORY

The words of Mercury are harsh after the songs of Apollo.

Love's Labour's Lost.

WE have seen that the swift-moving light-waves possess great advantages as a means of exploring the non-Euclidean property of space. But there is an old fable about the hare and the tortoise. The slow-moving planets have qualities which must not be overlooked. The light-wave traverses the region in a few minutes and makes its report; the planet plods on and on for centuries going over the same ground again and again. Each time it goes round it reveals a little about the space, and the knowledge slowly accumulates.

According to Newton's law a planet moves round the sun in an ellipse, and if there are no other planets disturbing it, the ellipse remains the same for ever. According to Einstein's law the path is very nearly an ellipse, but it does not quite close up; and in the next revolution the path has advanced slightly in the same direction as that in which the planet was moving. The orbit is thus an ellipse which very slowly revolves[1].

The exact prediction of Einstein's law is that in one revolution of the planet the orbit will advance through a fraction of a revolution equal to $3v^2/C^2$, where v is the speed of the planet and C the speed of light. The earth has $1/10,000$ of the speed of light; thus in one revolution (one year) the point where the earth is at greatest distance from the sun will move on $3/100,000,000$ of a revolution, or $0''.038$. We could not detect this difference in a year, but we can let it add up for a century at least. It would then be observable but for one thing—the earth's orbit is very blunt, very nearly circular, and so we cannot tell accurately enough which way it is pointing and how its sharpest apses move. We can choose a planet with higher speed so that the effect is increased, not only because v^2 is increased, but because the revolutions take less time; but, what is perhaps more important, we need a planet with a sharp elliptical orbit, so that it is easy to observe how its apses move round. Both these conditions are fulfilled in the case of Mercury. It is the fastest of the planets, and the predicted advance of the orbit amounts to $43''$ per century; further the eccentricity of its orbit is far greater than that of any of the other seven planets.

Now an unexplained advance of the orbit of Mercury had long been known. It had occupied the attention of Le Verrier, who, having successfully predicted the planet Neptune from the disturbances of Uranus, thought that the anomalous

[1]Appendix, Note 9.

motion of Mercury might be due to an interior planet, which was called Vulcan in anticipation. But, though thoroughly sought for, Vulcan has never turned up. Shortly before Einstein arrived at his law of gravitation, the accepted figures were as follows. The actual observed advance of the orbit was 574″ per century; the calculated perturbations produced by all the known planets amounted to 532″ per century. The excess of 42″ per century remained to be explained. Although the amount could scarcely be relied on to a second of arc, it was at least thirty times as great as the probable accidental error.

The big discrepancy from the Newtonian gravitational theory is thus in agreement with Einstein's prediction of an advance of 43″ per century.

The derivation of this prediction from Einstein's law can only be followed by mathematical analysis; but it may be remarked that any slight deviation from the inverse square law is likely to cause an advance or recession of the apse of the orbit. That a particle, if it does not move in a circle, should oscillate between two extreme distances is natural enough; it could scarcely do anything else unless it had sufficient speed to break away altogether. But the interval between the two extremes will not in general be half a revolution. It is only under the exact adjustment of the inverse square law that this happens, so that the orbit closes up and the next revolution starts at the same point. I do not think that any "simple explanation" of this property of the inverse-square law has been given; and it seems fair to remind those, who complain of the difficulty of understanding Einstein's prediction of the advance of the perihelion, that the real trouble is that they have not yet succeeded in making clear to the uninitiated this recondite result of the Newtonian theory. The slight modifications introduced by Einstein's law of gravitation upset this fine adjustment, so that the oscillation between the extremes occupies slightly more than a revolution. A simple example of this effect of a small deviation from the inverse-square law was actually given by Newton.

It had already been recognised that the change of mass with velocity may cause an advance of perihelion; but owing to the ambiguity of Newton's law of gravitation the discussion was unsatisfactory. It was, however, clear that the effect was too small to account for the motion of perihelion of Mercury, the prediction being $\frac{1}{2}v^2/C^2$, or at most v^2/C^2. Einstein's theory is the only one which gives the full amount $3v^2/C^2$.

It was suggested by Lodge that, this variation of mass with velocity might account for the whole motion of the orbit of Mercury, if account were taken of the sun's unknown absolute motion through the aether, combining sometimes additively and sometimes negatively with the orbital motion. In a discussion between him and the writer, it appeared that, if the absolute motion were sufficient to produce this effect on Mercury, it must give observable effects for Venus and the Earth; and these do not exist. Indeed from the close accordance of Venus and the Earth with observation, it is possible to conclude that, either the sun's motion through the aether is improbably small, or gravitation must conform to relativity, in the sense of the restricted principle (15), and conceal the effects of the increase of mass with speed so far as an additive uniform motion is concerned.

Unfortunately it is not possible to obtain any further test of Einstein's law of gravitation from the remaining planets. We have to pass over Venus and the

Earth, whose orbits are too nearly circular to show the advance of the apses observationally. Coming next to Mars with a moderately eccentric orbit, the speed is very much smaller, and the predicted advance is only $1''.3$ per century. Now the accepted figures show an observed advance (additional to that produced by known causes) of $5''$ per century, so that Einstein's correction improves the accordance of observation with theory; but, since the result for Mars is in any case scarcely trustworthy to $5''$ owing to the inevitable errors of observation, the improvement is not very important. The main conclusion is that Einstein's theory brings Mercury into line, without upsetting the existing good accordance of all the other planets.

We have tested Einstein's law of gravitation for fast movement (light) and for moderately slow movement (Mercury). For very slow movement it agrees with Newton's law, and the general accordance of the latter with observation can be transferred to Einstein's law. These tests appear to be sufficient to establish the law firmly. We can express it in this way.

Every particle or light-pulse moves so that the quantity s measured along its track between two points has the maximum possible value, where

$$ds^2 = -(1 - 2m/r)^{-1}\,dr^2 - r^2\,d\theta^2 + (1 - 2m/r)\,dt^2.$$

And the accuracy of the experimental test is sufficient to verify the coefficients as far as terms of order m/r in the coefficient of dr^2, and as far as terms of order m^2/r^2 in the coefficient of dt^2[2].

In this form the law appears to be firmly based on experiment, and the revision or even the complete abandonment of the general ideas of Einstein's theory would scarcely affect it.

These experimental proofs, that space in the gravitational field of the sun is non-Euclidean or curved, have appeared puzzling to those unfamiliar with the theory. It is pointed out that the experiments show that physical objects or loci are "warped" in the sun's field; but it is suggested that there is nothing to show that the space in which they exist is warped. The answer is that it does not seem possible to draw any distinction between the warping of physical space and the warping of physical objects which define space. If our purpose were merely to call attention to these phenomena of the gravitational field as curiosities, it would, no doubt, be preferable to avoid using words which are liable to be misconstrued. But if we wish to arrive at an understanding of the conditions of the gravitational field, we cannot throw over the vocabulary appropriate for that purpose, merely because there may be some who insist on investing the words with a metaphysical meaning which is clearly inappropriate to the discussion.

We come now to another kind of test. In the statement of the law of gravitation just given, a quantity s is mentioned; and, so far as that statement goes, s is merely an intermediary quantity defined mathematically. But in our theory we have been identifying s with interval-length, measured with an apparatus of scales and clocks; and it is very desirable to test whether this identification can be confirmed—whether the geometry of scales and clocks is the same as the geometry of moving particles and light-pulses.

---

[2] Appendix, Note 10.

The question has been mooted whether we may not divide the present theory into two parts. Can we not accept the law of gravitation in the form suggested above as a self-contained result proved by observation, leaving the further possibility that s is to be identified with interval-length open to debate? The motive is partly a desire to consolidate our gains, freeing them from the least taint of speculation; but perhaps also it is inspired by the wish to leave an opening by which clock-scale geometry, i.e. the space and time of ordinary perception, may remain Euclidean. Disregarding the connection of s with interval-length, there is no object in attributing any significance of length to it; it can be regarded as a dynamical quantity like Action, and the new law of gravitation can be expressed after the traditional manner without dragging in strange theories of space and time. Thus interpreted, the law perhaps loses its theoretical inevitability; but it remains strongly grounded on observation. Unfortunately for this proposal, it is impossible to make a clean division of the theory at the point suggested. Without some geometrical interpretation of s our conclusions as to the courses of planets and light-waves cannot be connected with the astronomical measurements which verify them. The track of a light-wave in terms of the coordinates r, θ, t cannot be tested directly; the coordinates afford only a temporary resting-place; and the measurement of the displacement of the star-image on the photographic plate involves a reconversion from the coordinates to s, which here appears in its significance as the interval in clock-scale geometry.

Thus even from the experimental standpoint, a rough correspondence of the quantity s occurring in the law of gravitation with the clock-scale interval is an essential feature. We have now to examine whether experimental evidence can be found as to the exactness of this correspondence.

It seems reasonable to suppose that a vibrating atom is an ideal type of clock. The beginning and end of a single vibration constitute two events, and the interval ds between two events is an absolute quantity independent of any mesh-system. This interval must be determined by the nature of the atom; and hence atoms which are absolutely similar will measure by their vibrations equal values of the absolute interval ds. Let us adopt the usual mesh-system (r, θ, t) for the solar system, so that

$$ds^2 = -\gamma^{-1}\, dr^2 - r^2\, d\theta^2 + \gamma\, dt^2.$$

Consider an atom momentarily at rest at some point in the solar system; we say *momentarily*, because it must undergo the acceleration of the gravitational field where it is. If ds corresponds to one vibration, then, since the atom has not moved, the corresponding dr and $d\theta$ will be zero, and we have

$$ds^2 = \gamma\, dt^2.$$

The *time* of vibration dt is thus $1/\sqrt{\gamma}$ times the *interval* of vibration ds.

Accordingly, if we have two similar atoms at rest at different points in the system, the interval of vibration will be the same for both; but the time of vibration will be proportional to the inverse square-root of γ, which differs for the two atoms. Since

$$\gamma = 1 - \frac{2m}{r}$$

$$1/\sqrt{\gamma} = 1 + \frac{m}{r}, \quad \text{very approximately.}$$

Take an atom on the surface of the sun, and a similar atom in a terrestrial laboratory. For the first, $1 + m/r = 1.00000212$, and for the second $1 + m/r$ is practically 1. The time of vibration of the solar atom is thus longer in the ratio 1.00000212, and it might be possible to test this by spectroscopic examination.

There is one important point to consider. The spectroscopic examination must take place in the terrestrial laboratory; and we have to test the period of the solar atom by the period of the waves emanating from it when they reach the earth. Will they carry the period to us unchanged? Clearly they must. The first and second pulse have to travel the same distance (r), and they travel with the same velocity (dr/dt); for the velocity of light in the mesh-system used is $1-2m/r$, and though this velocity depends on r, it does not depend on t. Hence the difference dt at one end of the waves is the same as that at the other end.

Thus in the laboratory the light from a solar source should be of greater period and greater wave-length (i.e. redder) than that from a corresponding terrestrial source. Taking blue light of wave-length 4000 Å, the solar lines should be displaced 4000.00000212, or 0.008 Å towards the red end of the spectrum.

The properties of a gravitational field of force are similar to those of a centrifugal field of force; and it may be of interest to see how a corresponding shift of the spectral lines occurs for an atom in a field of centrifugal force. Suppose that, as we rotate with the earth, we observe a very remote atom momentarily at rest relative to our rotating axes. The case is just similar to that of the solar atom; both are at rest relative to the respective mesh-systems; the solar atom is in a field of gravitational force, and the other is in a field of centrifugal force. The direction of the force is in both cases the same—from the earth towards the atom observed. Hence the atom in the centrifugal field ought also to vibrate more slowly, and show a displacement to the red in its spectral lines. It does, if the theory hitherto given is right. We can abolish the centrifugal force by choosing non-rotating axes. But the distant atom was at rest relative to the rotating axes, that is to say, it was whizzing round with them. Thus from the ordinary standpoint the atom has a large velocity relative to the observer, and, in accordance with Chapter 1, its vibrations slow down just as the aviator's watch did. The shift of spectral lines due to a field of centrifugal force is only another aspect of a phenomenon already discussed.

The expected shift of the spectral lines on the sun, compared with the corresponding terrestrial lines, has been looked for; but it has not been found.

In estimating the importance of this observational result in regard to the relativity theory, we must distinguish between a failure of the test and a definite conclusion that the lines are undisplaced. The chief investigators St John, Schwarzschild, Evershed, and Grebe and Bachem, seem to be agreed that the observed displacement is at any rate less than that predicted by the theory. The theory can therefore in no case claim support from the present evidence. But something more must be established, if the observations are to be regarded as in the slightest degree adverse to the theory. If for instance the mean deflection is found to be .004 instead of .008 Angström units, the only possible conclusion is that there are certain causes of displacement of the lines, acting in the solar atmosphere and not yet identified. No one could be much surprised if this were the case; and it would, of course, render the test nugatory. The case is not much altered if the observed displacement is .002 units, provided the latter quantity is above the accidental error of measurement; if we have to postulate some unexplained disturbance, it may just as well produce a displacement $-.006$ as $+.002$. For this reason Evershed's evidence is by no means adverse to the theory, since he finds unexplained displacements in any case. One set of lines measured by St John gave a mean displacement of .0036 units; and this also shows that the test has failed. The only evidence *adverse* to the theory, and not merely neutral, is a series of measures by St John on 17 cyanogen lines, which he regarded as most dependable. These gave a mean shift of exactly .000. If this stood alone we should certainly be disposed to infer that the test had gone against Einstein's theory, and that nothing had intervened to cast doubt on the validity of the test. The writer is unqualified to criticise these mutually contradictory spectroscopic conclusions; but he has formed the impression that the last-mentioned result obtained by St John has the greatest weight of any investigations up to the present[3].

It seems that judgment must be reserved; but it may be well to examine how the present theory would stand if the verdict of this third crucial experiment finally went against it.

It has become apparent that there is something illogical in the sequence we have followed in developing the theory, owing to the necessity of proceeding from the common ideas of space and time to the more fundamental properties of the absolute world. We started with a definition of the interval by measurements made with clocks and scales, and afterwards connected it with the tracks of moving particles. Clearly this is an inversion of the logical order. The simplest kind of clock is an elaborate mechanism, and a material scale is a very complex piece of apparatus. The best course then is to discover ds by exploration of space and time with a moving particle or light-pulse, rather than by measures with scales and clocks. On this basis by astronomical observation alone the formula for ds in the gravitational field of the sun has already been established. To proceed from this to determine exactly what is measured by a scale and a clock, it would at first seem necessary to have a detailed theory of the mechanisms involved in a scale and clock. But there is a short-cut which seems legitimate. This short-cut is in fact the Principle of Equivalence. Whatever the mechanism of the clock, whether it is a good clock or a bad clock, the intervals it is beating must be something absolute; the clock cannot know what mesh-system the observer is using, and therefore its absolute rate cannot be altered by position or motion which is relative merely to a mesh-system. Thus wherever it is placed, and however it moves, provided it is not constrained by impacts or electrical forces,

[3]A further paper by Grebe and Bachem (*Zeitschrift für Physik*, 1920, p. 51), received whilst this is passing through the press, makes out a case strongly favourable for the Einstein displacement, and reconciles the discordant results found by most of the investigators. But it may still be the best counsel to "wait and see," and I have made no alteration in the discussion here given.

it must always beat equal intervals as we have previously assumed. Thus a clock may fairly be used to measure intervals, even when the interval is defined in the new manner; any other result seems to postulate that it pays heed to some particular mesh-system[4].

Three modes of escape from this conclusion seem to be left open. A clock cannot pay any heed to the mesh-system used; but it may be affected by the kind of space-time around it[5]. The terrestrial atom is in a field of gravitation so weak that the space-time may be considered practically flat; but the space-time round the solar atom is not flat. It may happen that the two atoms actually detect this absolute difference in the world around them and do not vibrate with the same interval ds—contrary to our assumption above. Then the prediction of the shift of the lines in the solar spectrum is invalidated. Now it is very doubtful if an atom can detect the curving of the region it occupies, because curvature is only apparent when an extended region is considered; still an atom has some extension, and it is not impossible that its equations of motion involve the quantities $B^{\rho}_{\mu\nu\sigma}$ which distinguish gravitational from flat space-time. An apparently insuperable objection to this explanation is that the effect of curvature on the period would almost certainly be represented by terms of the form m^2/r^2, whereas to account for a negative result for the shift of the spectral lines terms of much greater order of magnitude m/r are needed.

The second possibility depends on the question whether it is possible for an atom at rest on the sun to be precisely similar to one on the earth. If an atom fell from the earth to the sun it would acquire a velocity of 610 km. per sec., and could only be brought to rest by a systematic hammering by other atoms. May not this have made a permanent alteration in its time-keeping properties? It is true that every atom is continually undergoing collisions, but it is just possible that the average solar atom has a different period from the average terrestrial atom owing to this systematic difference in its history.

What are the two events which mark the beginning and end of an atomic vibration? This question suggests a third possibility. If they are two absolute events, like the explosions of two detonators, then the interval between them will be a definite quantity, and our argument applies. But if, for example, an atomic vibration is determined by the revolution of an electron around a nucleus, it is not marked by any definite events. A revolution means a return to the same position as before; but we cannot define what is the same position as before without reference to some mesh-system. Hence it is not clear that there is any absolute interval corresponding to the vibration of an atom; an absolute interval only exists between two events absolutely defined.

It is unlikely that any of these three possibilities can negative the expected shift of the spectral lines. The uncertainties introduced by them are, so far as we can judge, of a much smaller order of magnitude. But it will be realised that this third test of Einstein's theory involves rather more complicated considerations than the two simple tests with light-waves and the moving planet. I think that

[4]Of course, there is always the possibility that this might be the case, though it seems unlikely. The essential point of the relativity theory is that (contrary to the common opinion) no experiments yet made have revealed any mesh-system of an absolute character, not that experiments never will reveal such a system.

[5]Appendix, Note 11.

a shift of the Fraunhofer lines is a highly probable prediction from the theory and I anticipate that experiment will ultimately confirm the prediction; but it is not entirely free from guess-work. These theoretical uncertainties are apart altogether from the great practical difficulties of the test, including the exact allowance for the unfamiliar circumstances of an absorbing atom in the sun's atmosphere.

Outside the three leading tests, there appears to be little chance of checking the theory unless our present methods of measurement are greatly improved. It is not practicable to measure the deflection of light by any body other than the sun. The apparent displacement of a star just grazing the limb of Jupiter should be $0''.017$. A hundredth of a second of arc is just about within reach of the most refined measurements with the largest telescopes. If the observation could be conducted under the same conditions as the best parallax measurements, the displacement could be detected but not measured with any accuracy. The glare from the light of the planet ruins any chance of success.

Most astronomers, who look into the subject, are entrapped sooner or later by a fallacy in connection with double stars. It is thought that when one component passes behind the other it will appear displaced from its true position, like a star passing behind the sun; if the size of the occulting star is comparable with that of the sun, the displacement should be of the same order, $1''.7$. This would cause a very conspicuous irregularity in the apparent orbit of a double star. But reference to 83 shows that an essential point in the argument was the enormous ratio of the distance QP of the star from the sun to the distance EF of the sun from the earth. It is only in these conditions that the apparent displacement of the object is equal to the deflection undergone by its light. It is easy to see that where this ratio is reversed, as in the case of the double star, the apparent displacement is an extremely small fraction of the deflection of the light. It would be quite imperceptible to observation.

If two independent stars are seen in the same line of vision within about $1''$, one being a great distance behind the other, the conditions seem at first more favourable. I do not know if any such pairs exist. It would seem that we ought to see the more distant star not only by the direct ray, which would be practically undisturbed, but also by a ray passing round the other side of the nearer star and bent by it to the necessary extent. The second image would, of course, be indistinguishable from that of the nearer star; but it would give it additional brightness, which would disappear in time when the two stars receded. But consider a pencil of light coming past the nearer star; the inner edge will be bent more than the outer edge, so that the divergence is increased. The increase is very small; but then the whole divergence of a pencil from a source some hundred billion miles away is very minute. It is easily calculated that the increased divergence would so weaken the light as to make it impossible to detect it when it reached us[6].

If two unconnected stars approached the line of sight still more closely, so that one almost occulted the other, observable effects might be perceived. When the proximity was such that the direct ray from the more distant star passed within about 100 million kilometres of the nearer star, it would begin to fade

[6]Appendix, Note 12.

appreciably. The course of the ray would not yet be appreciably deflected, but the divergence of the pencil would be rapidly increased, and less light from the star would enter our telescopes. The test is scarcely likely to be an important one, since a sufficiently close approach is not likely to occur; and in any case it would be difficult to feel confident that the fading was not due to a nebulous atmosphere around the nearer star.

The theory gives small corrections to the motion of the moon which have been investigated by de Sitter. Both the axis of the orbit and its line of intersection with the ecliptic should advance about $2''$ per century more than the Newtonian theory indicates. Neither observation nor Newtonian theory are as yet pushed to sufficient accuracy to test this; but a comparatively small increase in accuracy would make a comparison possible.

Since certain stars are perhaps ten times more massive than the sun, without the radius being unduly increased, they should show a greater shift of the spectral lines and might be more favourable for the third crucial test. Unfortunately the predicted shift is indistinguishable from that caused by a velocity of the star in the line-of-sight on Doppler's principle. Thus the expected shift on the sun is equivalent to that caused by a receding velocity of 0.634 kilometres per second. In the case of the sun we know by other evidence exactly what the line-of-sight velocity should be; but we have not this knowledge for other stars. The only indication that could be obtained would be the detection of an *average* motion of recession of the more massive stars. It seems rather unlikely that there should be a real preponderance of receding motions among stars taken indiscriminately from all parts of the sky; and the apparent effect might then be attributed to the Einstein shift. Actually the most massive stars (those of spectral type B) have been found to show an average velocity of recession of about 4.5 km. per sec., which would be explained if the values of m/r for these stars are about seven times greater than the value for the sun—a quite reasonable hypothesis. This phenomenon was well-known to astrophysicists some years before Einstein's theory was published. But there are so many possible interpretations that no stress should be placed on this evidence. Moreover the very diffuse "giant" stars of type M have also a considerable systematic velocity of recession, and for these m/r must be much less than for the sun.

9 Momentum and Energy

For spirits and men by different standards mete
The less and greater in the flow of time.
By sun and moon, primeval ordinances—
By stars which rise and set harmoniously—
By the recurring seasons, and the swing
This way and that of the suspended rod
Precise and punctual, men divide the hours,
Equal, continuous, for their common use.
Not so with us in the immaterial world;
But intervals in their succession
Are measured by the living thought alone
And grow or wane with its intensity.
And time is not a common property;
But what is long is short, and swift is slow
And near is distant, as received and grasped
By this mind and by that.

Newman, *Dream of Gerontius.*

One of the most important consequences of the relativity theory is the unification of inertia and gravitation.

The beginner in mechanics does not accept Newton's first law of motion without a feeling of hesitation. He readily agrees that a body at rest will remain at rest unless something causes it to move; but he is not satisfied that a body in motion will remain in uniform motion so long as it is not interfered with. It is quite natural to think that motion is an impulse which will exhaust itself, and that the body will finally come to a stop. The teacher easily disposes of the arguments urged in support of this view, pointing out the friction which has to be overcome when a train or a bicycle is kept moving uniformly. He shows that if the friction is diminished, as when a stone is projected across ice, the motion lasts for a longer time, so that if all interference by friction were removed uniform motion might continue indefinitely. But he glosses over the point that if there were no interference with the motion—if the ice were abolished altogether—the motion would be by no means uniform, but like that of a falling body. The teacher probably insists that the continuance of uniform motion does not require anything that can properly be called a *cause*. The property is given a name *inertia*; but it is thought of as an innate tendency in contrast to *force* which is an active cause. So long as forces are confined to the thrusts and tensions of elementary mechanics, where there is supposed to be direct contact of material, there is good ground for this distinction; we can visualise the active hammering of the molecules on the body, causing it to change its motion. But when force is extended to include the gravitational field the distinction is not so clear.

For our part we deny the distinction in this last case. Gravitational force is not an active agent working against the passive tendency of inertia. Gravitation and inertia are one. The uniform straight track is only relative to some mesh-system, which is assigned by arbitrary convention. We cannot imagine that a body looks round to see who is observing it and then feels an innate tendency to move in that observer's straight line—probably at the same time feeling an active force compelling it to move some other way. If there is anything that can be called an innate tendency it is the tendency to follow what we have called the natural track—the longest track between two points. We might restate the first law of motion in the form "Every body tends to move in the track in which it actually does move, except in so far as it is compelled by material impacts to follow some other track than that in which it would otherwise move." Probably no one will dispute this profound statement!

Whether the natural track is straight or curved, whether the motion is uniform or changing, a cause is in any case required. This cause is in all cases the combined inertia-gravitation. To have given it a name does not excuse us from attempting an explanation of it in due time. Meanwhile this identification of inertia and gravitation as arbitrary components of one property explains why weight is always proportional to inertia. This experimental fact verified to a very high degree of accuracy would otherwise have to be regarded as a remarkable law of nature.

We have learnt that the natural track is the longest track between two points; and since this is the only definable track having an absolute significance in nature, we seem to have a sufficient explanation of why an undisturbed particle must follow it. That is satisfactory, so far as it goes, but still we should naturally wish for a clearer picture of the cause—inertia-gravitation—which propels it in this track.

It has been seen that the gravitational field round a body involves a kind of curvature of space-time, and accordingly round each particle there is a minute pucker. Now at each successive instant a particle is displaced continuously in time if not in space; and so in our four-dimensional representation which gives a bird's-eye-view of all time, the pucker has the form of a long groove along the track of the particle. Now such a groove or pleat in a continuum cannot take an arbitrary course—as every dress-maker knows. Einstein's law of gravitation gives the rule according to which the curvatures at any point of space-time link on to those at surrounding points; so that when a groove is started in any direction the rest of its course can be forecasted. We have hitherto thought of the law of gravitation as showing how the pucker spreads out in space, cf. Newton's statement that the corresponding force weakens as the inverse square of the distance. But the law of Einstein equally shows how the gravitational field spreads out in time, since there is no absolute distinction of time and space. It can be deduced mathematically from Einstein's law that a pucker of the form corresponding to a particle necessarily runs along the track of greatest interval-length between two points.

The track of a particle of matter is thus determined by the interaction of the minute gravitational field, which surrounds and, so far as we know, constitutes it, with the general space-time of the region. The various forms which it can take, find their explanation in the new law of gravitation. The straight tracks of the stars and the curved tracks of the planets are placed on the same level, and receive the same kind of explanation. The one universal law, that the space-time continuum can be curved only in the first degree, is sufficient to prescribe the forms of all possible grooves crossing it.

The application of Einstein's law to trace the gravitational field not only through space but through time leads to a great unification of mechanics. If we have given for a start a narrow slice of space-time representing the state of the universe for a few seconds, with all the little puckers belonging to particles of matter properly described, then step by step all space-time can be linked on and the positions of the puckers shown at all subsequent times (electrical forces being excluded). Nothing is needed for this except the law of gravitation—that the curvature is only of the first degree—and there can thus be nothing in the predictions of mechanics which is not comprised in the law of gravitation. The conservation of mass, of energy, and of momentum must all be contained implicitly in Einstein's law.

It may seem strange that Einstein's law of gravitation should take over responsibility for the whole of mechanics; because in many mechanical problems gravitation in the ordinary sense can be neglected. But inertia and gravitation are unified; the law is also the law of inertia, and inertia or mass appears in all mechanical problems. When, as in many problems, we say that gravitation is negligible, we mean only that the interaction of the minute puckers with one another can be neglected; we do not mean that the interaction of the pucker of a particle with the general character of the space-time in which it lies can be neglected, because this constitutes the inertia of the particle.

The conservation of energy and the conservation of momentum in three independent directions, constitute together four laws or equations which are fundamental in all branches of mechanics. Although they apply when gravitation in the ordinary sense is not acting, they must be deducible like everything else in mechanics from the law of gravitation. It is a great triumph for Einstein's theory that his law gives correctly these experimental principles, which have generally been regarded as unconnected with gravitation. We cannot enter into the mathematical deduction of these equations; but we shall examine generally how they are arrived at.

It has already been explained that although the values of $G_{\mu\nu}$ are strictly zero everywhere in space-time, yet if we take average values through a small region containing a large number of particles of matter their average or "macroscopic" values will not be zero[1]. Expressions for these macroscopic values can be found in terms of the number, masses and motions of the particles. Since we have averaged the $G_{\mu\nu}$, we should also average the particles; that is to say, we replace

[1] It is the g's which are first averaged, then the $G_{\mu\nu}$ are calculated by the formulae in Note 5.

them by a distribution of continuous matter having equivalent properties. We thus obtain macroscopic equations of the form

$$G_{\mu\nu} = K_{\mu\nu},$$

where on the one side we have the somewhat abstruse quantities describing the kind of space-time, and on the other side we have well-known physical quantities describing the density, momentum, energy and internal stresses of the matter present. These macroscopic equations are obtained solely from the law of gravitation by the process of averaging.

By an exactly similar process we pass from Laplace's equation $\nabla^2\phi = 0$ to Poisson's equation for continuous matter $\nabla^2\phi = -4\pi\rho$, in the Newtonian theory of gravitation.

When continuous matter is admitted, *any* kind of space-time becomes possible. The law of gravitation instead of denying the possibility of certain kinds, states what values of $K_{\mu\nu}$, i.e. what distribution and motion of continuous matter in the region, are a necessary accompaniment. This is no contradiction with the original statement of the law, since that referred to the case in which continuous matter is denied or excluded. Any set of values of the potentials is now possible; we have only to calculate by the formulae the corresponding values of $G_{\mu\nu}$, and we at once obtain ten equations giving the $K_{\mu\nu}$ which define the conditions of the matter necessary to produce these potentials. But suppose the necessary distribution of matter through space and time is an impossible one, violating the laws of mechanics! No, there is only one law of mechanics, the law of gravitation; we have specified the distribution of matter so as to satisfy $G_{\mu\nu} = K_{\mu\nu}$, and there can be no other condition for it to fulfil. The distribution must be mechanically possible; it might, however, be unrealisable in practice, involving inordinately high or even negative density of matter.

In connection with the law for empty space, $G_{\mu\nu} = 0$, it was noticed that whereas this apparently forms a set of ten equations, only six of them can be independent. This was because ten equations would suffice to determine the ten potentials precisely, and so fix not only the kind of space-time but the mesh-system. It is clear that we must preserve the right to draw the mesh-system as we please; it is fixed by arbitrary choice not by a law of nature. To allow for the four-fold arbitrariness of choice, there must be four relations always satisfied by the $G_{\mu\nu}$, so that when six of the equations are given the remaining four become tautological.

These relations must be identities implied in the mathematical definition of $G_{\mu\nu}$; that is to say, when the $G_{\mu\nu}$ have been written out in full according to their definition, and the operations indicated by the identities carried out, all the terms will cancel, leaving only $0 = 0$. The essential point is that the four relations follow from the mode of formation of the $G_{\mu\nu}$ from their simpler constituents ($g_{\mu\nu}$ and their differential coefficients) and apply universally. These four identical relations have actually been discovered[2].

When in continuous matter $G_{\mu\nu} = K_{\mu\nu}$ clearly the same four relations must exist between the $K_{\mu\nu}$, not now as identities, but as consequences of the law of gravitation, viz. the equality of $G_{\mu\nu}$ and $K_{\mu\nu}$.

[2] Appendix, Note 13.

Thus the four dimensions of the world bring about a four-fold arbitrariness of choice of mesh-system; this in turn necessitates four identical relations between the $G_{\mu\nu}$; and finally, in consequence of the law of gravitation, these identities reveal four new facts or laws relating to the density, energy, momentum or stress of matter, summarised in the expressions $K_{\mu\nu}$.

These four laws turn out to be the laws of conservation of momentum and energy.

The argument is so general that we can even assert that corresponding to any *absolute* property of a volume of a world of four dimensions (in this case, *curvature*), there must be four *relative* properties which are conserved. This might be made the starting-point of a general inquiry into the necessary qualities of a permanent perceptual world, i.e. a world whose substance is conserved.

There is another law of physics which was formerly regarded as fundamental—the conservation of mass. Modern progress has somewhat altered our position with regard to it; not that its validity is denied, but it has been reinterpreted, and has finally become merged in the conservation of energy. It will be desirable to consider this in detail.

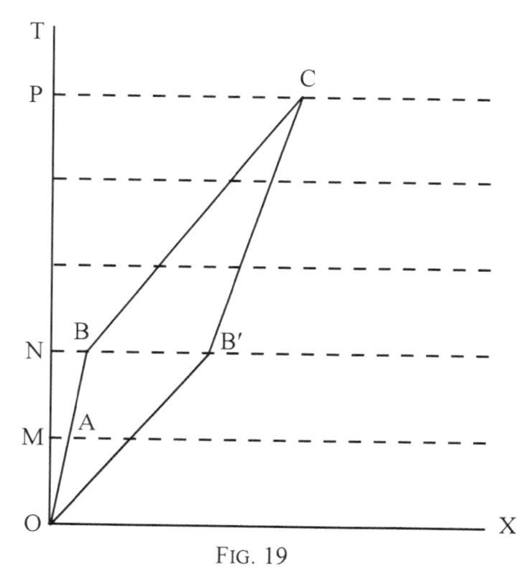

FIG. 19

It was formerly supposed that the mass of a particle was a number attached to the particle, expressing an intrinsic property, which remained unaltered in all its vicissitudes. If m is this number, and u the velocity of the particle, the momentum is mu; and it is through this relation, coupled with the law of conservation of momentum that the mass m was defined. Let us take for example two particles of masses $m_1 = 2$ and $m_2 = 3$, moving in the same straight line. In the space-time diagram for an observer S the velocity of the first particle will be represented by a direction OA (Fig. 19). The first particle moves through a space MA in unit time, so that MA is equal to its velocity referred to the observer S. Prolonging the line OA to meet the second time-partition, NB is equal to the velocity multiplied by the mass 2; thus the horizontal distance NB represents the momentum. Similarly, starting from B and drawing BC in the direction of the velocity of m_2, prolonged through three time-partitions, the horizontal progress from B represents the momentum of the second particle. The length PC then represents the total momentum of the system of two particles.

Suppose that some change of their velocities occurs, not involving any transference of momentum from outside, e.g. a collision. Since the total momentum PC is unaltered, a similar construction made with the new velocities must again bring us to C; that is to say, the new velocities are represented by the directions OB', $B'C$, where B' is some other point on the line NB.

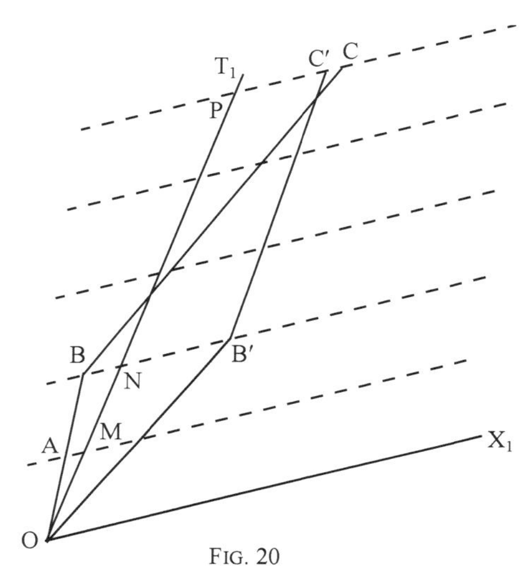

FIG. 20

Now examine how this will appear to some other observer S_1 in uniform motion relative to S. His transformation of space and time has been described in Chapter 3 and is represented in Fig. 20, which shows how his time-partitions run as compared with those of S. The same actual motion is, of course, represented by parallel directions in the two diagrams; but the interpretation as a velocity MA is different in the two cases. Carrying the velocity of m_1 through two time-partitions, and of m_2 through three time-partitions, as before, we find that the total momentum for the observer S_1 is represented by PC (Fig. 20); but making a similar construction with the velocities after collision, we arrive at a different point C'. Thus whilst momentum is conserved for the observer S, it has altered from PC to PC' for the observer S_1.

The discrepancy arises because in the construction the lines are prolonged to meet partitions which are different for the two observers. The rule for determining momentum ought to be such that both observers make the same construction, independent of their partitions, so that both arrive by the two routes at the same point C. Then it will not matter if, through their different measures of time, one observer measures momentum by horizontal progress and the other by oblique progress; both will agree that the momentum has not been altered by the collision. To describe such a construction, we must use the interval which is alike for both observers; make the interval-length of OB equal to 2 units, and that of BC equal to 3 units, disregarding the mesh-system altogether. Then both observers will make the same diagram and arrive at the same point C (different from C or C' in the previous diagrams). Then if momentum is conserved for one observer, it will be conserved for the other.

This involves a modified definition of momentum. Momentum must now be the mass multiplied by the change of position δx per lapse of interval δs, instead of per lapse of time δt. Thus

$$\text{momentum} = m\frac{\delta x}{\delta s}$$

$$\text{instead of momentum} = m\frac{\delta x}{\delta t},$$

and the mass m still preserves its character as an invariant number associated with the particle.

Whether the momentum as now defined is actually conserved or not, is a matter for experiment, or for theoretical deduction from the law of gravitation. The point is that with the original definition general conservation is impossible, because if it held good for one observer it could not hold for another. The new definition makes general conservation possible. Actually this form of the momentum is the one deduced from the law of gravitation through the identities already described. With regard to experimental confirmation it is sufficient at present to state that in all ordinary cases the interval and the time are so nearly equal that such experimental foundation as existed for the law of conservation of the old momentum is just as applicable to the new momentum.

Thus in the theory of relativity momentum appears as an invariant mass multiplied by a modified velocity $\delta x/\delta s$. The physicist, however, prefers for practical purposes to keep to the old definition of momentum as mass multiplied by the velocity $\delta x/\delta t$. We have

$$m\frac{\delta x}{\delta s} = m\frac{\delta t}{\delta s}\frac{\delta x}{\delta t},$$

accordingly the momentum is separated into two factors, the velocity $\delta x/\delta t$, and a mass $M = m\delta t/\delta s$, which is no longer an invariant for the particle but depends on its motion relative to the observer's space and time. In accordance with the usual practice of physicists the mass (unless otherwise qualified) is taken to mean the quantity M.

Using unaccelerated rectangular axes, we have by definition of s

$$\delta s^2 = \delta t^2 - \delta x^2 - \delta y^2 - \delta z^2,$$

so that

$$\begin{aligned}\left(\frac{\delta s}{\delta t}\right)^2 &= 1 - \left(\frac{\delta x}{\delta t}\right)^2 - \left(\frac{\delta y}{\delta t}\right)^2 - \left(\frac{\delta z}{\delta t}\right)^2, \\ &= 1 - u^2,\end{aligned}$$

where u is the resultant velocity of the particle (the velocity of light being unity). Hence

$$M = \frac{m}{\sqrt{(1-u^2)}}.$$

Thus the mass increases as the velocity increases, the factor being the same as that which determines the FitzGerald contraction.

The increase of mass with velocity is a property which challenges experimental test. For success it is necessary to be able to experiment with high velocities and to apply a known force large enough to produce appreciable deflection in the fast-moving particle. These conditions are conveniently fulfilled by the small negatively charged particles emitted by radio-active substances, known as β particles, or the similar particles which constitute cathode rays. They attain speeds up to 0.8 of the velocity of light, for which the increase of mass is in the ratio 1.66; and the negative charge enables a large electric or magnetic force to be applied. Modern experiments fully confirm the theoretical increase of mass, and show that the factor $1/\sqrt{(1-u^2)}$ is at least approximately correct. The experiment was originally performed by Kaufmann; but much greater accuracy has been obtained by recent modified methods.

Unless the velocity is very great the mass M may be written

$$m/\sqrt{(1-u^2)} = m + \tfrac{1}{2}mu^2.$$

Thus it consists of two parts, the mass when at rest, together with the second term which is simply the energy of the motion. If we can say that the term m represents a kind of potential energy concealed in the matter, mass can be identified with energy. The increase of mass with velocity simply means that the energy of motion has been added on. We are emboldened to do this because in the case of an electrical charge the electrical mass is simply the energy of the static field. Similarly the mass of light is simply the electromagnetic energy of the light.

In our ordinary units the velocity of light is not unity, and a rather artificial distinction between mass and energy is introduced. They are measured by different units, and energy E has a mass E/C^2 where C is the velocity of light in the units used. But it seems very probable that mass and energy are two ways of measuring what is essentially the same thing, in the same sense that the parallax and distance of a star are two ways of expressing the same property of location. If it is objected that they ought not to be confused inasmuch as they are distinct properties, it must be pointed out that they are not sense-properties, but mathematical terms expressing the dividend and product of more immediately apprehensible properties, viz. momentum and velocity. They are essentially mathematical compositions, and are at the disposal of the mathematician.

This proof of the variation of mass with velocity is much more general than that based on the electrical theory of inertia. It applies immediately to matter in bulk. The masses m_1 and m_2 need not be particles; they can be bodies of any size or composition. On the electrical theory alone, there is no means of deducing the variation of mass of a planet from that of an electron.

It has to be remarked that, although the inertial mass of a particle only comes under physical measurement in connection with a change of its motion, it is just when the motion is changing that the conception of its mass is least definite; because it is at that time that the kinetic energy, which forms part of the mass, is being passed on to another particle or radiated into the surrounding field; and it is scarcely possible to define the moment at which this energy ceases to be associated with the particle and must be reckoned as broken loose. The amount of energy or mass in a given region is always a definite quantity; but the amount attributable to a particle is only definite when the motion is uniform. In rigorous work it is generally necessary to consider the mass not of a particle but of a region.

The motion of matter from one place to another causes an alteration of the gravitational field in the surrounding space. If the motion is uniform, the field is simply convected; but if the motion is accelerated, something of the nature of a gravitational wave is propagated outwards. The velocity of propagation is the velocity of light. The exact laws are not very simple because we have seen that the gravitational field modifies the velocity of light; and so the disturbance itself modifies the velocity with which it is propagated. In the same way the exact laws of propagation of sound are highly complicated, because the disturbance of the air by sound modifies the speed with which it is propagated. But the approximate laws of propagation of gravitation are quite simple and are the same as those of electromagnetic disturbances.

After mass and energy there is one physical quantity which plays a very fundamental part in modern physics, known as *Action*. *Action* here is a very technical term, and is not to be confused with Newton's "Action and Reaction." In the relativity theory in particular this seems in many respects to be the most fundamental thing of all. The reason is not difficult to see. If we wish to speak of the continuous matter present *at* any particular point of space and time, we must use the term *density*. Density multiplied by volume in space gives us *mass* or, what appears to be the same thing, *energy*. But from our space-time point of view, a far more important thing is density multiplied by a four-dimensional volume of space and time; this is *action*. The multiplication by three dimensions gives mass or energy; and the fourth multiplication gives mass or energy multiplied by time. Action is thus mass multiplied by time, or energy multiplied by time, and is more fundamental than either.

Action is the curvature of the world. It is scarcely possible to visualise this statement, because our notion of curvature is derived from surfaces of two dimensions in a three-dimensional space, and this gives too limited an idea of the possibilities of a four-dimensional surface in space of five or more dimensions. In two dimensions there is just one total curvature, and if that vanishes the surface is flat or at least can be unrolled into a plane. In four dimensions there are many coefficients of curvature; but there is one curvature *par excellence*, which is, of course, an invariant independent of our mesh-system. It is the quantity we have denoted by G. It does not follow that if the curvature vanishes space-time is flat; we have seen in fact that in a natural gravitational field space-time is not flat although there may be no mass or energy and therefore no action or curvature.

Wherever there is matter[3] there is action and therefore curvature; and it is interesting to notice that in ordinary matter the curvature of the space-time world is by no means insignificant. For example, in water of ordinary density the curvature is the same as that of space in the form of a sphere of radius 570,000,000 kilometres. The result is even more surprising if expressed in time

[3]It is rather curious that there is no action in space containing only light. Light has mass (M) of the ordinary kind; but the invariant mass (m) vanishes.

units; the radius is about half-an-hour.

It is difficult to picture quite what this means; but at least we can predict that a globe of water of 570, 000, 000 km. radius would have extraordinary properties. Presumably there must be an upper limit to the possible size of a globe of water. So far as I can make out a homogeneous mass of water of about this size (and no larger) could exist. It would have no centre, and no boundary, every point of it being in the same position with respect to the whole mass as every other point of it—like points on the *surface* of a sphere with respect to the surface. Any ray of light after travelling for an hour or two would come back to the starting point. Nothing could enter or leave the mass, because there is no boundary to enter or leave by; in fact, it is coextensive with space. There could not be any other world anywhere else, because there isn't an "anywhere else."

The mass of this volume of water is not so great as the most moderate estimates of the mass of the stellar system. Some physicists have predicted a distant future when all energy will be degraded, and the stellar universe will gradually fall together into one mass. Perhaps then these strange conditions will be realised!

The law of gravitation, the laws of mechanics, and the laws of the electromagnetic field have all been summed up in a single Principle of Least Action. For the most part this unification was accomplished before the advent of the relativity theory, and it is only the addition of gravitation to the scheme which is novel. We can see now that if action is something absolute, a configuration giving minimum action is capable of absolute definition; and accordingly we should expect that the laws of the world would be expressible in some such form. The argument is similar to that by which we first identified the natural tracks of particles with the tracks of greatest interval-length. The fact that some such form of law is inevitable, rather discourages us from seeking in it any clue to the structural details of our world.

Action is one of the two terms in pre-relativity physics which survive unmodified in a description of the absolute world. The only other survival is entropy. The coming theory of relativity had cast its shadow before; and physics was already converging to two great generalisations, the principle of least action and the second law of thermodynamics or principle of maximum entropy.

We are about to pass on to recent and more shadowy developments of this subject; and this is an appropriate place to glance back on the chief results that have emerged. The following summary will recall some of the salient points.

1. The order of events in the external world is a four-dimensional order.

2. The observer either intuitively or deliberately constructs a system of meshes (space and time partitions) and locates the events with respect to these.

3. Although it seems to be theoretically possible to describe phenomena without reference to any mesh-system (by a catalogue of coincidences), such a description would be cumbersome. In practice, physics describes the relations of the events to our mesh-system; and all the terms of elementary physics and of daily life refer to this relative aspect of the world.

4. Quantities like length, duration, mass, force, etc. have no absolute significance; their values will depend on the mesh-system to which they are referred. When this fact is realised, the results of modern experiments relating to changes of length of rigid bodies are no longer paradoxical.

5. There is no fundamental mesh-system. In particular problems, and more particularly in restricted regions, it may be possible to choose a mesh-system which follows more or less closely the lines of absolute structure in the world, and so simplify the phenomena which are related to it. But the world-structure is not of a kind which can be traced in an exact way by mesh-systems, and in any large region the mesh-system drawn must be considered arbitrary. In any case the systems used in current physics are arbitrary.

6. The study of the absolute structure of the world is based on the "interval" between two events close together, which is an absolute attribute of the events independent of any mesh-system. A world-geometry is constructed by adopting the interval as the analogue of distance in ordinary geometry.

7. This world-geometry has a property unlike that of Euclidean geometry in that the interval between two real events may be real or imaginary. The necessity for a physical distinction, corresponding to the mathematical distinction between real and imaginary intervals, introduces us to the separation of the four-dimensional order into time and space. But this separation is not unique, and the separation commonly adopted depends on the observer's track through the four-dimensional world.

8. The geodesic, or track of maximum or minimum interval-length between two distant events, has an absolute significance. And since no other kind of track can be defined absolutely, it is concluded that the tracks of freely moving particles are geodesics.

9. In Euclidean geometry the geodesics are straight lines. It is evidently impossible to choose space and time-reckoning so that all free particles in the solar system move in straight lines. Hence the geometry must be non-Euclidean in a field of gravitation.

10. Since the tracks of particles in a gravitational field are evidently governed by some law, the possible geometries must be limited to certain types.

11. The limitation concerns the absolute structure of the world, and must be independent of the choice of mesh-system. This narrows down the possible discriminating characters. Practically the only reasonable suggestion is that the world must (in empty space) be "curved no higher than the first degree"; and this is taken as the law of gravitation.

12. The simplest type of hummock with this limited curvature has been investigated. It has a kind of infinite chimney at the summit, which we must suppose cut out and filled up with a region where this law is not obeyed, i.e. with a particle of matter.

13. The tracks of the geodesics on the hummock are such as to give a very close accordance with the tracks computed by Newton's law of gravitation. The slight differences from the Newtonian law have been experimentally verified by the motion of Mercury and the deflection of light.

14. The hummock might more properly be described as a ridge extending linearly. Since the interval-length along it is real or time-like, the ridge can be taken as a time-direction. Matter has thus a continued existence in time. Further, in order to conform with the law, a small ridge must always follow a geodesic in the general field of space-time, confirming the conclusion arrived at under (8).

15. The laws of conservation of energy and momentum in mechanics can be deduced from this law of world-curvature.

16. Certain phenomena such as the FitzGerald contraction and the variation of mass with velocity, which were formerly thought to depend on the behaviour of electrical forces concerned, are now seen to be general consequences of the relativity of knowledge. That is to say, length and mass being the relations of some absolute thing to the observer's mesh-system, we can foretell how these relations will be altered when referred to another mesh-system.

10 Towards Infinity

The geometer of to-day knows nothing about the nature of actually existing space at an infinite distance; he knows nothing about the properties of this present space in a past or a future eternity. He knows, indeed, that the laws assumed by Euclid are true with an accuracy that no direct experiment can approach, not only in this place where we are, but in places at a distance from us that no astronomer has conceived; but he knows this as of Here and Now; beyond his range is a There and Then of which he knows nothing at present, but may ultimately come to know more.

W. K. Clifford (1873).

The great stumbling-block for a philosophy which denies absolute space is the experimental detection of absolute rotation. The belief that the earth rotates on its axis was suggested by the diurnal motions of the heavenly bodies; this observation is essentially one of relative rotation, and, if the matter rested there, no difficulty would be felt. But we can detect the same rotation, or a rotation very closely equal to it, by methods which do not seem to bring the heavenly bodies into consideration; and such a rotation is apparently absolute. The planet Jupiter is covered with cloud, so that an inhabitant would probably be unaware of the existence of bodies outside; yet he could quite well measure the rotation of Jupiter. By the gyrocompass he would fix two points on the planet—the north and south poles. Then by Foucault's experiment on the change of the plane of motion of a freely suspended pendulum, he would determine an angular velocity about the poles. Thus there is certainly a definite physical constant, an angular velocity about an axis, which has a fundamental importance for the inhabitants of Jupiter; the only question is whether we are right in giving it the name absolute rotation.

Contrast this with absolute translation. Here it is not a question of giving the right name to a physical constant; the inhabitants of Jupiter would find no constant to name. We see at once that a relativity theory of translation is on a different footing from a relativity theory of rotation. The duty of the former is to explain facts; the duty of the latter is to explain away facts.

Our present theory seems to make a start at tackling this problem, but gives it up. It permits the observer, if he wishes, to consider the earth as non-rotating, but surrounded by a field of centrifugal force; all the other bodies in the universe are then revolving round the earth in orbits mainly controlled by this field of centrifugal force. Astronomy on this basis is a little cumbersome; but all the phenomena are explained perfectly. The centrifugal force is part of the gravitational field, and obeys Einstein's law of gravitation, so that the laws of nature are completely satisfied by this representation. One awkward question remains, What causes the centrifugal force? Certainly not the earth which is here represented as non-rotating. As we go further into space to look for a cause, the centrifugal force becomes greater and greater, so that the more we defer the debt the heavier the payment demanded in the end. Our present theory is like the debtor who does not mind how big an obligation accumulates satisfied that he can always put off the payment. It chases the cause away to infinity, content that the laws of nature—the relations between contiguous parts of the world—are satisfied all the way.

One suggested loophole must be explored. Our new law of gravitation admits that a rapid motion of the attracting body will affect the field of force. If the earth is non-rotating, the stars must be going round it with terrific speed. May they not in virtue of their high velocities produce gravitationally a sensible field of force on the earth, which we recognise as the centrifugal field? This would be a genuine elimination of absolute rotation, attributing all effects indifferently to the rotation of the earth the stars being at rest, or to the revolution of the stars the earth being at rest; nothing matters except the relative rotation. I doubt whether anyone will persuade himself that the stars have anything to do with the phenomenon. We do not believe that if the heavenly bodies were all annihilated it would upset the gyrocompass. In any case, precise calculation shows that the centrifugal force could not be produced by the motion of the stars, so far as they are known.

We are therefore forced to give up the idea that the signs of the earth's rotation—the protuberance of its equator, the phenomena of the gyrocompass, etc.—are due to a rotation relative to any matter we can recognise. The philosopher who persists that a rotation which is not relative to matter is unthinkable, will no doubt reply that the rotation must then be relative to some matter which we have not yet recognised. We have hitherto been greatly indebted to the suggestions of philosophy in evolving this theory, because the suggestions related to the things we know about; and, as it turned out, they were confirmed by experiment. But as physicists we cannot take the same interest in the new demand; we do not necessarily challenge it, but it is outside our concern. Physics demands of its scheme of nature something else besides truth, namely a certain quality that we may call convergence. The law of conservation of energy is only strictly true when the whole universe is taken into account; but its value in physics lies in the fact that it is *approximately* true for a very limited system. Physics is an exact science because the chief essentials of a problem are limited to a few conditions; and it draws near to the truth with ever-increasing approximation as it widens its purview. The approximations of physics form a convergent series. History, on the other hand, is very often like a divergent series; no approximation to its course is reached until the last term of the infinite series has been included in the data of prediction. Physics, if it wishes to retain its advantage, must take its own course, formulating those laws which are approximately true for the limited data of sense, and extending them into the unknown. The relativity of rotation is not approximately true for the data of sense, although it may possibly be true when the unknown as well as the known are included.

The same considerations that apply to rotation apply to acceleration, although the difficulty is less striking. We can if we like attribute to the sun some arbitrary acceleration, balancing it by introducing a uniform gravitational field. Owing to this field the rest of the stars will move with the same acceleration and no phenomena will be altered. But then it seems necessary to find a cause for this field. It is not produced by the gravitation of the stars. Our only course is to pursue the cause further and further towards infinity; the further we put it away, the greater the mass of attracting matter needed to produce it. On the other hand, the earth's absolute acceleration does not intrude on our attention in the way that its absolute rotation does[1].

We are vaguely conscious of a difficulty in these results; but if we examine it closely, the difficulty does not seem to be a very serious one. The theory of relativity, as we have understood it, asserts that our partitions of space and time are introduced by the observer and are irrelevant to the laws of nature; and therefore the current quantities of physics, length, duration, mass, force, etc., which are relative to these partitions, are not things having an absolute significance in nature. But we have never denied that there are features of the world having an absolute significance; in fact, we have spent much time in finding such features. The geodesics or natural tracks have been shown to have an absolute significance; and it is possible in a limited region of the world to choose space and time partitions such that all geodesics become approximately straight lines. We may call this a "natural" frame for that region, although it is not as a rule the space and time adopted in practice; it is for example the space and time of the observers in the falling projectile, not of Newton's super-observer. It is capable of absolute definition, except that it is ambiguous in regard to uniform motion. Now the rotation of the earth determined by Foucault's pendulum experiment is the rotation referred to this natural frame. But we must have misunderstood our own theory of relativity altogether, if we think there is anything inadmissible in an absolute rotation of such a kind.

Material particles and geodesics are both features of the absolute structure of the world; and a rotation relative to geodesic structure does not seem to be on any different footing from a velocity relative to matter. There is, however, the striking feature that rotation seems to be relative not merely to the local geodesic structure but to a generally accepted universal frame; whereas it is necessary to specify precisely what matter a velocity is measured with respect to. This is largely a question of how much accuracy is needed in specifying velocities and rotations, respectively. If in stating the speed of a β particle we do not mind an error of $10,000$ kilometres a second, we need not specify precisely what star or planet its velocity is referred to. The moon's (local) angular velocity is sometimes given to fourteen significant figures; I doubt if any universal frame is well-defined enough for this accuracy. There is no doubt much greater continuity in the geodesic structure in different parts of the world than in the material structure; but the difference is in degree rather than in principle.

It is probable that here we part company from many of the continental rela-

[1]To determine even roughly the earth's absolute acceleration we should need a fairly full knowledge of the disturbing effects of all the matter in the universe. A similar knowledge would be required to determine the absolute rotation *accurately*; but all the matter likely to exist would have so small an effect, that we can at once assume that the absolute rotation is very nearly the same as the experimentally determined rotation.

tivists, who give prominent place to a principle known as the *law of causality*—that only those things are to be regarded as being in causal connection which are capable of being actually observed. This seems to be interpreted as placing matter on a plane above geodesic structure in regard to the formulation of physical laws, though it is not easy to see in what sense a distribution of matter can be regarded as more observable than the field of influence in surrounding space which makes us aware of its existence. The principle itself is debateable; that which is observable to us is determined by the accident of our own structure, and the law of causality seems to impose our own limitations on the free interplay of entities in the world outside us. In this book the tradition of Faraday and Maxwell still rules our outlook; and for us matter and electricity are but incidental points of complexity, the activity of nature being primarily in the so-called empty spaces between.

The vague universal frame to which rotation is referred is called the *inertial frame*. It is definite in the flat space-time far away from all matter. In the undulating country corresponding to the stellar universe it is not a precise conception; it is rather a rude outline, arbitrary within reasonable limits, but with the general course indicated. The reason for the term inertial frame is of interest. We can quite freely use a mesh-system deviating widely from the inertial frame (e.g. rotating axes); but we have seen that there is a postponed debt to pay in the shape of an apparently uncaused field of force. But is there no debt to pay, even when the inertial frame is used? In that case there is no gravitational or centrifugal force at infinity; but there is still inertia, which is of the same nature. The distinction between force as requiring a cause and inertia as requiring no cause cannot be sustained. We shall not become any more solvent by commuting our debt into pure inertia. The debt is inevitable whatever mesh-system is used; we are only allowed to choose the form it shall take.

The debt after all is a very harmless one. At infinity we have the absolute geodesics in space-time, and we have our own arbitrarily drawn mesh-system. The relation of the geodesics to the mesh-system decides whether our axes shall be termed rotating or non-rotating; and ideally it is this relation that is determined when a so-called absolute rotation is measured. No one could reasonably expect that there would be no determinable relation. On the other hand uniform translation does not affect the relation of the geodesics to the mesh-system—if they were straight lines originally, they remain straight lines—thus uniform translation cannot be measured except relative to matter.

We have been supposing that the conditions found in the remotest parts of space accessible to observation can be extrapolated to infinity; and that there are still definite natural tracks in space-time far beyond the influence of matter. Feelings of objection to this view arise in certain minds. It is urged that as matter influences the course of geodesics it may well be responsible for them altogether; so that a region outside the field of action of matter could have no geodesics, and consequently no intervals. All the potentials would then necessarily be zero. Various modified forms of this objection arise; but the main feeling seems to be that it is unsatisfactory to have certain conditions prevailing in the world, which can be traced away to infinity and so have, as it were, their source at infinity; and there is a desire to find some explanation of the inertial frame as built up through conditions at a finite distance.

Now if all intervals vanished space-time would shrink to a point. Then there would be no space, no time, no inertia, no anything. Thus a cause which creates intervals and geodesics must, so to speak, extend the world. We can imagine the world stretched out like a plane sheet; but then the stretching cause—the cause of the intervals—is relegated beyond the bounds of space and time, i.e. to infinity. This is the view objected to, though the writer does not consider that the objection has much force. An alternative way is to inflate the world from inside, as a balloon is blown out. In this case the stretching force is not relegated to infinity, and ruled outside the scope of experiment; it is acting at every point of space and time, curving the world to a sphere. We thus get the idea that space-time may have an essential curvature on a great scale independent of the small hummocks due to recognised matter.

It is not necessary to speculate whether the curvature is produced (as in the balloon) by some pressure applied from a fifth dimension. For us it will appear as an innate tendency of four-dimensional space-time to curve. It may be asked, what have we gained by substituting a natural curvature of space-time for a natural stretched condition corresponding to the inertial frame? As an explanation, nothing. But there is this difference, that the theory of the inertial frame can now be included in the differential law of gravitation instead of remaining outside and additional to the law.

It will be remembered that one clue by which we previously reached the law of gravitation was that flat space-time must be compatible with it. But if space-time is to have a small natural curvature independent of matter this condition is now altered. It is not difficult to find the necessary alteration of the law[2]. It will contain an additional, and at present unknown, constant, which determines the size of the world.

Spherical space is not very easy to imagine. We have to think of the properties of the surface of a sphere—the two-dimensional case—and try to conceive something similar applied to three-dimensional space. Stationing ourselves at a point let us draw a series of spheres of successively greater radii. The surface of a sphere of radius r should be proportional to r^2; but in spherical space the areas of the more distant spheres begin to fall below the proper proportion. There is not so much room out there as we expected to find. Ultimately we reach a sphere of biggest possible area, and beyond it the areas begin to decrease[3]. The last sphere of all shrinks to a point—our antipodes. Is there nothing beyond this? Is there a kind of boundary there? There is nothing beyond and yet there is no boundary. On the earth's surface there is nothing beyond our own antipodes but there is no boundary there.

The difficulty is that we try to realise this spherical world by imagining how it would appear to us and to our measurements. There has been nothing in our experience to compare it with, and it seems fantastic. But if we could get rid of the personal point of view, and regard the sphericity of the world as a statement of the type of order of events outside us, we should think that it was a simple and natural order which is as likely as any other to occur in the world.

---

[2]Appendix, Note 14.

[3]The area is, of course, to be determined by measurement of some kind.

In such a world there is no difficulty about accumulated debt at the boundary. There is no boundary. The centrifugal force increases until we reach the sphere of greatest area, and then, still obeying the law of gravitation, diminishes to zero at the antipodes. The debt has paid itself automatically.

We must not exaggerate what has been accomplished by this modification of the theory. A new constant has been introduced into the law of gravitation which gives the world a definite extension. Previously there was nothing to fix the scale of the world; it was simply given *a priori* that it was infinite. Granted extension, so that the intervals are not invariably zero, we can determine geodesics everywhere, and hence mark out the inertial frame.

Spherical space-time, that is to say a four-dimensional continuum of space and imaginary time forming the surface of a sphere in five dimensions, has been investigated by Prof. de Sitter. If real time is used the world is spherical in its space dimensions, but open towards plus and minus infinity in its time dimension, like an hyperboloid. This happily relieves us of the necessity of supposing that as we progress in time we shall ultimately come back to the instant we started from! History never repeats itself. But in the space dimensions we should, if we went on, ultimately come back to the starting point. This would have interesting physical results, and we shall see presently that Einstein has a theory of the world in which the return can actually happen; but in de Sitter's theory it is rather an abstraction, because, as he says, "all the paradoxical phenomena can only happen after the end or before the beginning of eternity."

The reason is this. Owing to curvature in the time dimension, as we examine the condition of things further and further from our starting point, our time begins to run faster and faster, or to put it another way natural phenomena and natural clocks slow down. The condition becomes like that described in Mr H. G. Wells's story "The new accelerator."

When we reach half-way to the antipodal point, time stands still. Like the Mad Hatter's tea party, it is always 6 o'clock; and nothing whatever can happen however long we wait. There is no possibility of getting any further, because everything including light has come to rest here. All that lies beyond is for ever cut off from us by this barrier of time; and light can never complete its voyage round the world.

That is what happens when the world is viewed from one station; but if attracted by such a delightful prospect, we proceeded to visit this scene of repose, we should be disappointed. We should find nature there as active as ever. We thought time was standing still, but it was really proceeding there at the usual rate, as if in a fifth dimension of which we had no cognisance. Casting an eye back on our old home we should see that time apparently had stopped still there. Time in the two places is proceeding in directions at right angles, so that the progress of time at one point has no relation to the perception of time at the other point. The reader will easily see that a being confined to the surface of a sphere and not cognisant of a third dimension, will, so to speak, lose one of his dimensions altogether when he watches things occurring at a point 90 away. He regains it if he visits the spot and so adapts himself to the two dimensions which prevail there.

It might seem that this kind of fantastic world-building can have little to do with practical problems. But that is not quite certain. May we not be able actually to observe the slowing down of natural phenomena at great distances from us? The most remote objects known are the spiral nebulae, whose distances may perhaps be of the order a million light years. If natural phenomena are slowed down there, the vibrations of an atom are slower, and its characteristic spectral lines will appear displaced to the red. We should generally interpret this as a Doppler effect, implying that the nebula is receding. The motions in the line-of-sight of a number of nebulae have been determined, chiefly by Prof. Slipher. The data are not so ample as we should like; but there is no doubt that large receding motions greatly preponderate. This may be a genuine phenomenon in the evolution of the material universe; but it is also possible that the interpretation of spectral displacement as a receding velocity is erroneous; and the effect is really the slowing of atomic vibrations predicted by de Sitter's theory.

Prof. Einstein himself prefers a different theory of curved space-time. His world is cylindrical—curved in the three space dimensions and straight in the time dimension. Since time is no longer curved, the slowing of phenomena at great distances from the observer disappears, and with it the slight experimental support given to the theory by the observations of spiral nebulae. There is no longer a barrier of eternal rest, and a ray of light is able to go round the world.

In various ways crude estimates of the size of the world both on de Sitter's and Einstein's hypotheses have been made; and in both cases the radius is thought to be of the order 10^{13} times the distance of the earth from the sun. A ray of light from the sun would thus take about 1000 million years to go round the world; and after the journey the rays would converge again at the starting point, and then diverge for the next circuit. The convergent would have all the characteristics of a real sun so far as light and heat are concerned, only there would be no substantial body present. Thus corresponding to the sun we might see a series of ghosts occupying the positions where the sun was 1000, 2000, 3000, etc., million years ago, if (as seems probable) the sun has been luminous for so long.

It is rather a pleasing speculation that records of the previous states of the sidereal universe may be automatically reforming themselves on the original sites. Perhaps one or more of the many spiral nebulae are really phantoms of our own stellar system. Or it may be that only a proportion of the stars are substantial bodies; the remainder are optical ghosts revisiting their old haunts. It is, however, unlikely that the light rays after their long journey would converge with the accuracy which this theory would require. The minute deflections by the various gravitational fields encountered on the way would turn them aside, and the focus would be blurred. Moreover there is a likelihood that the light would gradually be absorbed or scattered by matter diffused in space, which is encountered on the long journey.

It is sometimes suggested that the return of the light-wave to its starting point can most easily be regarded as due to the force of gravitation, there being sufficient mass distributed through the universe to control its path in a closed orbit. We should have no objection in principle to this way of looking at it; but we doubt whether it is correct in fact. It is quite possible for light to return to its starting point in a world without gravitation. We can roll flat space-time into a cylinder and join the edges; its geometry will still be Euclidean and there will be no gravitation; but a ray of light can go right round the cylinder and return to the starting point in space. Similarly in Einstein's more complex type of cylinder (three dimensions curved and one dimension linear), it seems likely that the return of the light is due as much to the connectivity of his space, as to the non-Euclidean properties which express the gravitational field.

For Einstein's cylindrical world it is necessary to postulate the existence of vast quantities of matter (not needed on de Sitter's theory) far in excess of what has been revealed by our telescopes. This additional material may either be in the form of distant stars and galaxies beyond our limits of vision, or it may be uniformly spread through space and escape notice by its low density. There is a definite relation between the average density of matter and the radius of the world; the greater the radius the smaller must be the average density.

Two objections to this theory may be urged. In the first place, absolute space and time are restored for phenomena on a cosmical scale. The ghost of a star appears at the spot where the star was a certain number of million years ago; and from the ghost to the present position of the star is a definite distance—the absolute motion of the star in the meantime[4]. The world taken as a whole has one direction in which it is not curved; that direction gives a kind of absolute time distinct from space. Relativity is reduced to a local phenomenon; and although this is quite sufficient for the theory hitherto described, we are inclined to look on the limitation rather grudgingly. But we have already urged that the relativity theory is not concerned to deny the possibility of an absolute time, but to deny that it is concerned in any experimental knowledge yet found; and it need not perturb us if the conception of absolute time turns up in a new form in a theory of phenomena on a cosmical scale, as to which no experimental knowledge is yet available. Just as each limited observer has his own particular separation of space and time, so a being coextensive with the world might well have a special separation of space and time natural to him. It is the time for this being that is here dignified by the title "absolute."

Secondly, the revised law of gravitation involves a new constant which depends on the total amount of matter in the world; or conversely the total amount of matter in the world is determined by the law of gravitation. This seems very hard to accept—at any rate without some plausible explanation of how the adjustment is brought about. We can see that, the constant in the law of gravitation being fixed, there may be some upper limit to the amount of matter possible; as more and more matter is added in the distant parts, space curves round and ultimately closes; the process of adding more matter must stop, because there is no more space, and we can only return to the region already dealt with. But there seems nothing to prevent a defect of matter, leaving space unclosed. Some mechanism seems to be needed, whereby either gravitation creates matter, or all the matter in the universe conspires to define a law of gravitation.

Although this appears to the writer rather bewildering, it is welcomed by those philosophers who follow the lead of Mach. For it leads to the result that

[4]The ghost is not formed where the star is now. If two stars were near together when the light left them their ghosts must be near together, although the stars may now be widely separated.

the extension of space and time depends on the amount of matter in the world—partly by its direct effect on the curvature and partly by its influence on the constant of the law of gravitation. The more matter there is, the more space is created to contain it, and if there were no matter the world would shrink to a point.

In the philosophy of Mach a world without *matter* is unthinkable. Matter in Mach's philosophy is not merely required as a test body to display properties of something already there, which have no physical meaning except in relation to matter; it is an essential factor in causing those properties which it is able to display. Inertia, for example, would not appear by the insertion of one test body in the world; in some way the presence of other matter is a necessary condition. It will be seen how welcome to such a philosophy is the theory that space and the inertial frame come into being with matter, and grow as it grows. Since the laws of inertia are part of the law of gravitation, Mach's philosophy was summed up—perhaps unconsciously—in the profound saying "If there were no matter in the universe, the law of gravitation would fall to the ground."

No doubt a world without matter, in which nothing could ever happen, would be very uninteresting; and some might deny its claim to be regarded as a world at all. But a world uniformly filled with matter would be equally dull and unprofitable; so there seems to be little object in denying the possibility of the former and leaving the latter possible.

The position can be summed up as follows:—in a space without absolute features, an absolute rotation would be as meaningless as an absolute translation; accordingly, the existence of an experimentally determined quantity generally identified with absolute rotation requires explanation. It was remarked on p. 30 that it would be difficult to devise a plan of the world according to which uniform motion has no significance but non-uniform motion is significant; but such a world has been arrived at—a plenum, of which the absolute features are intervals and geodesics. In a limited region this plenum gives a natural frame with respect to which an acceleration or rotation (but not a velocity) capable of absolute definition can be measured. In the case of rotation the local distortions of the frame are of comparatively little account; and this explains why in practice rotation appears to have reference to some world-wide inertial frame.

Thus absolute rotation does not indicate any logical flaw in the theory hitherto developed; and there is no need to accept any modification of our views. Possibly there may be a still wider relativity theory, in which our supposed plenum is to be regarded as itself an abstraction of the relations of the matter distributed throughout the world, and not existent apart from such matter. This seems to exalt matter rather unnecessarily. It may be true; but we feel no necessity for it, unless experiment points that way. It is with some such underlying idea that Einstein's cylindrical space-time was suggested, since this cannot exist without matter to keep it stretched. Now we freely admit that our assumption of perfect flatness in the remote parts of space was arbitrary, and there is no justification for insisting on it. A small curvature is possible both conceptually and experimentally. The arguments on both sides have hitherto been little more than prejudices, which would be dissipated by any experimental or theoretical lead in one direction. Weyl's theory of the electromagnetic field, discussed in the next chapter, assigns a definite function to the curvature of space; and this considerably alters the aspect of the question. We are scarcely sufficiently advanced to offer a final opinion; but the conception of cylindrical space-time seems to be favoured by this new development of the theory.

Some may be inclined to challenge the right of the Einstein theory, at least as interpreted in this book, to be called a relativity theory. Perhaps it has not all the characteristics which have at one time or another been associated with that name; but the reader, who has followed us so far, will see how our search for an absolute world has been guided by a recognition of the relativity of the measurements of physics. It may be urged that our geodesics ought not to be regarded as fundamental; a geodesic has no meaning in itself; what we are really concerned with is the relation of a particle following a geodesic to all the other matter of the world and the geodesic cannot be thought of apart from such other matter. We would reply, "Your particle of matter is not fundamental; it has no meaning in itself; what you are really concerned with is its 'field'—the relation of the geodesics about it to the other geodesics in the world—and matter cannot be thought of apart from its field." It is all a tangle of relations; physical theory starts with the simplest constituents, philosophical theory with the most familiar constituents. They may reach the same goal; but their methods are often incompatible.

11 Electricity and Gravitation

Thou shalt not have in thy bag divers weights, a great and a small. Thou shalt not have in thine house divers measures, a great and a small. But thou shalt have a perfect and just weight, a perfect and just measure shalt thou have.

Book of Deuteronomy.

The relativity theory deduces from geometrical principles the existence of gravitation and the laws of mechanics of matter. Mechanics is derived from geometry, not by *adding* arbitrary hypotheses, but by *removing* unnecessary assumptions, so that a geometer like Riemann might almost have foreseen the more important features of the actual world. But nature has in reserve one great surprise—electricity.

Electrical phenomena are not in any way a misfit in the relativity theory, and historically it is through them that it has been developed. Yet we cannot rest satisfied until a deeper unity between the gravitational and electrical properties of the world is apparent. The electron, which seems to be the smallest particle of matter, is a singularity in the gravitational field and also a singularity in the electrical field. How can these two facts be connected? The gravitational field is the expression of some state of the world, which also manifests itself in the natural geometry determined with measuring appliances; the electric field must also express some state of the world, but we have not as yet connected it with natural geometry. May there not still be unnecessary assumptions to be removed, so that a yet more comprehensive geometry can be found, in which gravitational and electrical fields both have their place?

There *is* an arbitrary assumption in our geometry up to this point, which it is desirable now to point out. We have based everything on the "interval," which, it has been said, is something which all observers, whatever their motion or whatever their mesh-system, can measure absolutely, agreeing on the result. This assumes that they are provided with identical standards of measurement—scales and clocks. But if A is in motion relative to B and wishes to hand his standards to B to check his measures, he must stop their motion; this means in practice that he must bombard his standards with material molecules until they come to rest. Is it fair to assume that no alteration of the standard is caused by this process? Or if A measures time by the vibrations of a hydrogen atom, and space by the wave-length of the vibration, still it is necessary to stop the atom by a collision in which electrical forces are involved?

The standard of length in physics is the length in the year 1799 of a bar deposited at Paris. Obviously no interval is ever compared directly with that length; there must be a continuous chain of intermediate steps extending like a geodetic triangulation through space and time, first along the past history of the scale actually used, then through intermediate standards, and finally along the history of the Paris metre itself. It may be that these intermediate steps are of no importance—that the same result is reached by whatever route we approach the standard; but clearly we ought not to make that assumption without due consideration. We ought to construct our geometry in such a way as to show that there are intermediate steps, and that the comparison of the interval with the ultimate standard is not a kind of action at a distance.

To compare intervals in different directions at a point in space and time does not require this comparison with a distant standard. The physicist's method of describing phenomena near a point P is to lay down for comparison (1) a mesh-system, (2) a unit of length (some kind of material standard), which can also be used for measuring time, the velocity of light being unity. With this system of reference he can measure in terms of his unit small intervals PP' running in any direction from P, summarising the results in the fundamental formula

$$ds^2 = g_{11}\, dx_1^2 + g_{22}\, dx_2^2 + \cdots + 2g_{12}\, dx_1 dx_2 + \cdots .$$

If now he wishes to measure intervals near a distant point Q, he must lay down a mesh-system and a unit of measure there. He naturally tries to simplify matters by using what he would call the *same* unit of measure at P and Q, either by transporting a material rod or some equivalent device. If it is immaterial by what route the unit is carried from P to Q, and replicas of the unit carried by different routes all agree on arrival at Q, this method is at any rate explicit. The question whether the unit at Q defined in this way is *really* the same as that at P is mere metaphysics. But if the units carried by different routes disagree, there is no unambiguous means of identifying a unit at Q with the unit at P. Suppose P is an event at Cambridge on March 1, and Q at London on May 1; we are contemplating the possibility that there will be a difference in the results of measures made with our standard in London on May 1, according as the standard is taken up to London on March 1 and remains there, or is left at Cambridge and taken up on May 1. This seems at first very improbable; but our reasons for allowing for this possibility will appear presently. If there is this ambiguity the only possible course is to lay down (1) a mesh-system filling all the space and time considered, (2) a definite unit of interval, or gauge, *at every point of space and time.* The geometry of the world referred to such a system will be more complicated than that of Riemann hitherto used; and we shall see that it is necessary to specify not only the 10 g's, but four other functions of position, which will be found to have an important physical meaning.

The observer will naturally simplify things by making the units of gauge at different points as nearly as possible equal, judged by ordinary comparisons. But the fact remains that, when the comparison depends on the route taken, exact equality is not definable; and we have therefore to admit that the *exact* standards are laid down at every point independently.

It is the same problem over again as occurs in regard to mesh-systems. We lay down particular rectangular axes near a point P; presently we make some observations near a distant point Q. To what coordinates shall the latter be referred? The natural answer is that we must use the same coordinates as we were using at P. But, except in the particular case of flat space, there is no means of defining exactly what coordinates at Q are the *same* as those at P. In many cases the ambiguity may be too trifling to trouble us; but in exact work the only course is to lay down a definite mesh-system extending throughout space, the precise route of the partitions being necessarily arbitrary. We now find that we have to add to this by placing in each mesh a gauge whose precise length must be arbitrary. Having done this the next step is to make measurements of intervals (using our gauges). This connects the absolute properties of the world with our arbitrarily drawn mesh-system and gauge-system. And so by measurement we determine the g's and the new additional quantities, which determine the geometry of our chosen system of reference, and at the same time contain within themselves the absolute geometry of the world—the kind of space-time which exists in the field of our experiments.

Having laid down a unit-gauge at every point, we can speak quite definitely of the change in interval-length of a measuring-rod moved from point to point, meaning, of course, the change compared with the unit-gauges. Let us take a rod of interval-length l at P, and move it successively through the displacements dx_1, dx_2, dx_3, dx_4; and let the result be to increase its length in terms of the gauges by the amount λl. The change depends as much on the difference of the gauges at the two points as on the behaviour of the rod; but there is no possibility of separating the two factors. It is clear that λ will not depend on l, because the change of length must be proportional to the original length—unless indeed our whole idea of measurement by comparison with a gauge is wrong[1]. Further it will not depend on the direction of the rod either in its initial or final positions because the interval-length is independent of direction. (Of course, the space-length would change, but that is already taken care of by the g's.) λ can thus only depend on the displacements dx_1, dx_2, dx_3, dx_4, and we may write it

$$\lambda = \kappa_1\, dx_1 + \kappa_2\, dx_2 + \kappa_3\, dx_3 + \kappa_4\, dx_4,$$

so long as the displacements are small. The coefficients κ_1, κ_2, κ_3, κ_4 apply to the neighbourhood of P, and will in general be different in different parts of space.

This indeed assumes that the result is independent of the order of the displacements dx_1, dx_2, dx_3, dx_4—that is to say that the ambiguity of the comparison by different routes disappears in the limit when the whole route is sufficiently small. It is parallel with our previous implicit assumption that although the length of the track from a point P to a distant point Q depends on the route, and no definite meaning can be attached to the interval between them without specifying a route, yet in the limit there is a definite small interval between P and Q when they are sufficiently close together.

To understand the meaning of these new coefficients κ let us briefly recapitulate what we understand by the g's. Primarily they are quantities derived from

---

[1] We refuse to contemplate the idea that when the metre rod changes its length to two metres, each centimetre of it changes to three centimetres.

experimental measurements of intervals, and describe the geometry of the space and time partitions which the observer has chosen. As consequential properties they describe the field of force, gravitational, centrifugal, etc., with which he perceives himself surrounded. They relate to the particular mesh-system of the observer; and by altering his mesh-system, he can alter their values, though not entirely at will. From their values can be deduced intrinsic properties of the world—the *kind* of space-time in which the phenomena occur. Further they satisfy a definite condition—the law of gravitation—so that not all mathematically possible space-times and not all arbitrary values of the g's are such as can occur in nature.

All this applies equally to the κ's, if we substitute gauge-system for mesh-system, and some at present unknown force for gravitation. They can theoretically be determined by interval-measurement; but they will be more conspicuously manifested to the observer through their consequential property of describing some kind of field of force surrounding him. The κ's refer to the arbitrary gauge-system of the observer; but he cannot by altering his gauge-system alter their values entirely at will. Intrinsic properties of the world are contained in their values, unaffected by any change of gauge-system. Further we may expect that they will have to satisfy some law corresponding to the law of gravitation, so that not all arbitrary values of the κ's are such as can occur in nature.

It is evident that the κ's must refer to some type of phenomenon which has not hitherto appeared in our discussion; and the obvious suggestion is that they refer to the electromagnetic field. This hypothesis is strengthened when we recall that the electromagnetic field is, in fact, specified at every point by the values of four quantities, viz. the three components of electromagnetic vector potential, and the scalar potential of electrostatics. Surely it is more than a coincidence that the physicist needs just four more quantities to specify the state of the world at a point in space, and four more quantities are provided by removing a rather illogical restriction on our system of geometry of natural measures.

[The general reader will perhaps pardon a few words addressed especially to the mathematical physicist. Taking the ordinary unaccelerated rectangular coordinates x, y, z, t, let us write F, G, H, $-\Phi$ for κ_1, κ_2, κ_3, κ_4, then

$$\frac{dl}{l} = \lambda = F\,dx + G\,dy + H\,dx - \Phi\,dt.$$

From which, by integration,

$$\log l + \text{const.} = \int (F\,dx + G\,dy + H\,dz - \Phi\,dt).$$

The length l will be independent of the route taken if

$$F\,dx + G\,dy + H\,dz - \Phi\,dt$$

is a perfect differential. The condition for this is

$$\frac{\partial H}{\partial y} - \frac{\partial G}{\partial z} = 0, \qquad \frac{\partial F}{\partial z} - \frac{\partial H}{\partial x} = 0, \qquad \frac{\partial G}{\partial x} - \frac{\partial F}{\partial y} = 0,$$

$$-\frac{\partial \Phi}{\partial x} - \frac{\partial F}{\partial t} = 0, \qquad -\frac{\partial \Phi}{\partial y} - \frac{\partial G}{\partial t} = 0, \qquad -\frac{\partial \Phi}{\partial z} - \frac{\partial H}{\partial t} = 0.$$

If F, G, H, Φ are the potentials of electromagnetic theory, these are precisely the expressions for the three components of magnetic force and the three components of electric force, given in the text-books. Thus the condition that distant intervals can be compared directly without specifying a particular route of comparison is that the electric and magnetic forces are zero in the intervening space and time.

It may be noted that, even when the coordinate system has been defined, the electromagnetic potentials are not unique in value; but arbitrary additions can be made provided these additions form a perfect differential. It is just this flexibility which in our geometrical theory appears in the form of the arbitrary choice of gauge-system. The electromagnetic *forces* on the other hand are independent of the gauge-system, which is eliminated by "curling."]

It thus appears that the four new quantities appearing in our extended geometry may actually be the four potentials of electromagnetic theory; and further, when there is no electromagnetic field our previous geometry is valid. But in the more general case we have to adopt the more general geometry in which there appear fourteen coefficients, ten describing the gravitational and four the electrical conditions of the world.

We ought now to seek the law of the electromagnetic field on the same lines as we sought for the law of gravitation, laying down the condition that it must be independent of mesh-system and gauge-system since it seeks to limit the possible kinds of world which can exist in nature. Happily this presents no difficulty, because the law expressed by Maxwell's equations, and universally adopted, fulfils the conditions. There is no need to modify it fundamentally as we modified the law of gravitation. We do, however, generalise it so that it applies when a gravitational field is present at the same time—not merely, as given by Maxwell, for flat space-time. The deflection of electromagnetic waves (light) by a gravitational field is duly contained in this generalised law.

Strictly speaking the laws of gravitation and of the electromagnetic field are not two laws but one law, as the geometry of the g's and the κ's is one geometry. Although it is often convenient to separate them, they are really parts of the general condition limiting the possible kinds of metric that can occur in empty space.

It will be remembered that the four-fold arbitrariness of our mesh-system involved four identities, which were found to express the conservation of energy and momentum. In the new geometry there is a fifth arbitrariness, namely that of the selected gauge-system. This must also give rise to an identity; and it is found that the new identity expresses the law of conservation of electric charge.

A grasp of the new geometry may perhaps be assisted by a further comparison. Suppose an observer has laid down a line of a certain length and in a certain direction at a point P, and he wishes to lay down an exactly similar line at a distant point Q. If he is in flat space there will be no difficulty; he will have to proceed by steps, a kind of triangulation, but the route chosen is of no importance. We know definitely that there is just one direction at Q parallel to the original direction at P; and it is in ordinary geometry supposed that the length is equally determinate. But if space is not flat the case is different. Imagine a two-dimensional observer confined to the curved surface of the earth trying to perform this task. As he does not appreciate the third dimension he will not immediately perceive the impossibility; but he will find that the direction which he has transferred to Q differs according to the route chosen. Or if he went round a complete circuit he would find on arriving back at P that the direction he had so carefully tried to preserve on the journey did not agree with that originally drawn[2]. We describe this by saying that in curved space, direction is not integrable; and it is this non-integrability of direction which characterises the gravitational field. In the case considered the length would be preserved throughout the circuit; but it is possible to conceive a more general kind of space in which the length which it was attempted to preserve throughout the circuit, as well as the direction, disagreed on return to the starting point with that originally drawn. In that case length is not integrable; and the non-integrability of length characterises the electromagnetic field. Length associated with direction is called a vector; and the combined gravitational and electric field describe that influence of the world on our measurements by which a vector carried by physical measurement round a closed circuit changes insensibly into a different vector.

The welding together of electricity and gravitation into one geometry is the work of Prof. H. Weyl, first published in 1918[3]. It appears to the writer to carry conviction, although up to the present no experimental test has been proposed. It need scarcely be said that the inconsistency of length for an ordinary circuit would be extremely minute[4], and the ordinary manifestations of the electromagnetic field are the consequential results of changes which would be imperceptible to direct measurement. It will be remembered that the gravitational field is likewise perceived by the consequential effects, and not by direct interval-measurement.

But the theory does appear to require that, for example, the time of vibration of an atom is not quite independent of its previous history. It may be assumed that the previous histories of terrestrial atoms are so much alike that there are no significant differences in their periods. The possibility that the systematic difference of history of solar and terrestrial atoms may have an effect on the expected shift of the spectral lines on the sun has already been alluded to. It seems doubtful, however, whether the effect could attain the necessary magnitude.

It may seem difficult to identify these abstract geometrical qualities of the world with the physical forces of electricity and magnetism. How, for instance, can the change in the length of a rod taken round a circuit in space and time be responsible for the sensations of an electric shock? The geometrical potentials (κ) obey the recognised laws of electromagnetic potentials, and each entity in the physical theory—charge, electric force, magnetic element, light, etc.—has its exact analogue in the geometrical theory; but is this formal correspondence a sufficient ground for identification? The doubt which arises in our minds is due to a failure to recognise the formalism of all physical knowledge. The suggestion

[2] It might be thought that if the observer preserved mentally the original direction in three-dimensional space, and obtained the direction at any point in the two-dimensional space by projecting it, there would be no ambiguity. But the three-dimensional space in which a curved two-dimensional space is conceived to exist is quite arbitrary. A two-dimensional observer cannot ascertain by any observation whether he is on a plane or a cylinder, a sphere or any other convex surface of the same total curvature.

[3] Appendix, Note 15.

[4] I do not think that any numerical estimate has been made.

"This is not the thing I am speaking of, though it behaves exactly like it in all respects" carries no physical meaning. Anything which behaves exactly like electricity must manifest itself to us as electricity. Distinction of form is the only distinction that physics can recognise; and distinction of individuality, if it has any meaning at all, has no bearing on physical manifestations.

We can only explore the world with apparatus, which is itself part of the world. Our idealised apparatus is reduced to a few simple types—a neutral particle, a charged particle, a rigid scale, etc. The absolute constituents of the world are related in various ways, which we have studied, to the indications of these test-bodies. The main features of the absolute world are so simple that there is a redundancy of apparatus at our disposal; and probably all that there is to be known could theoretically be found out by exploration with an uncharged particle. Actually we prefer to look at the world as revealed by exploration with scales and clocks—the former for measuring so-called imaginary intervals, and the latter for real intervals; this gives us a unified geometrical conception of the world. Presumably, we could obtain a unified mechanical conception by taking the moving uncharged particle as standard indicator; or a unified electrical conception by taking the charged particle. For particular purposes one test-body is generally better adapted than others. The gravitational field is more sensitively explored with a moving particle than a scale. Although the electrical field can theoretically be explored by the change of length of a scale taken round a circuit, a far more sensitive way is to use a little bit of the scale—an electron. And in general for practical efficiency, we do not use any simple type of apparatus, but a complicated construction built up with a view to a particular experiment. The reason for emphasising the theoretical interchangeability of test-bodies is that it brings out the unity and simplicity of the world; and for that reason there is an importance in characterising the electromagnetic condition of the world by reference to the indications of a scale and clock, however inappropriate they may be as practical test-bodies.

Weyl's theory opens up interesting avenues for development. The details of the further steps involve difficult mathematics; but a general outline is possible. As on Einstein's more limited theory there is at any point an important property of the world called the curvature; but on the new theory it is not an absolute quantity in the strictest sense of the word. It is independent of the observer's mesh-system, but it depends on his gauge. It is obvious that the number expressing the radius of curvature of the world at a point must depend on the unit of length; so we cannot say that the curvatures at two points are absolutely equal, because they depend on the gauges assigned at the two points. Conversely the radius of curvature of the world provides a natural and absolute gauge at every point; and it will presumably introduce the greatest possible symmetry into our laws if the observer chooses this, or some definite fraction of it, as his gauge. He, so to speak, forces the world to be spherical by adopting at every point a unit of length which will make it so. Actual rods as they are moved about change their lengths compared with this absolute unit according to the route taken, and the differences correspond to the electromagnetic field. Einstein's curved space appears in a perfectly natural manner in this theory; no part of space-time is flat, even in the absence of ordinary matter, for that would mean infinite radius of curvature, and there would be no natural gauge to determine, for example, the dimensions of an electron—the electron could not know how large it ought to be, unless it had something to measure itself against.

The connection between the form of the law of gravitation and the total amount of matter in the world now appears less mysterious. The curvature of space indirectly provides the gauge which we use for measuring the amount of matter in the world.

Since the curvature is not independent of the gauge, Weyl does not identify it with the most fundamental quantity in nature. There is, however, a slightly more complicated invariant which is a pure number, and this is taken to be Action[5]. We can thus mark out a definite volume of space and time, and say that the action within it is 5, without troubling to define coordinates or the unit of measure! It might be expected that the action represented by the number 1 would have specially interesting properties; it might, for instance, be an atom of action and indivisible. Experiment has isolated what are believed to be units of action, which at least in many phenomena behave as indivisible atoms called quanta; but the theory, as at present developed, does not permit us to represent the quantum of action by the number 1. The quantum is a very minute fraction of the absolute unit.

When we come across a pure number having some absolute significance in the world it is natural to speculate on its possible interpretation. It might represent a number of discrete entities; but in that case it must necessarily be an integer, and it seems clear that action can have fractional values. An angle is commonly represented as a pure number, but it has not really this character; an angle can only be measured in terms of a unit of angle, just as a length is measured in terms of a unit of length. I can only think of one interpretation of a fractional number which can have an absolute significance, though doubtless there are others. The number may represent the *probability* of something, or some function of a probability. The precise function is easily found. We combine probabilities by multiplying, but we combine the actions in two regions by adding; hence the logarithm of a probability is indicated. Further, since the logarithm of a probability is necessarily negative, we may identify action provisionally with minus the logarithm of the statistical probability of the state of the world which exists.

The suggestion is particularly attractive because the Principle of Least Action now becomes the Principle of Greatest Probability. The law of nature is that the actual state of the world is that which is statistically most probable.

Weyl's theory also shows that the mass of a portion of matter is necessarily positive; on the original theory no adequate reason is given why negative matter should not exist. It is further claimed that the theory shows to some extent why the world is four-dimensional. To the mathematician it seems so easy to generalise geometry to n dimensions, that we naturally expect a world of four dimensions to have an analogue in five dimensions. Apparently this is not the case, and there are some essential properties, without which it could scarcely be a world, which exist only for four dimensions. Perhaps this may be compared with the well-known difficulty of generalising the idea of a knot; a knot can exist

[5] Appendix, Note 16.

in space of any odd number of dimensions, but not in space of an even number.

Finally the theory suggests a mode of attacking the problem of how the electric charge of an electron is held together; at least it gives an explanation of why the gravitational force is so extremely weak compared with the electric force. It will be remembered that associated with the mass of the sun is a certain length, called the gravitational mass, which is equal to 1×5 kilometres. In the same way the gravitational mass or radius of an electron is 7×10^{-56} cms. Its electrical properties are similarly associated with a length 2×10^{-13} cms., which is called the electrical radius. The latter is generally supposed to correspond to the electron's actual dimensions. The theory suggests that the ratio of the gravitational to the electrical radius, 3×10^{42}, ought to be of the same order as the ratio of the latter to the radius of curvature of the world. This would require the radius of space to be of the order 6×10^{29} cms., or 2×10^{11} parsecs., which though somewhat larger than the provisional estimates made by de Sitter, is within the realm of possibility.

12 On the Nature of Things

Hippolyta	This is the silliest stuff that ever I heard.
Theseus	The best in this kind are but shadows; and the worst are no worse, if imagination amend them.

A Midsummer-Night's Dream.

THE constructive results of the theory of relativity are based on two principles which have been enunciated—the restricted principle of relativity, and the principle of equivalence. These may be summed up in the statement that uniform motion and fields of force are purely relative. In their more formal enunciations they are experimental generalisations, which can be admitted or denied; if admitted, all the observational results obtained by us can be deduced mathematically without any reference to the views of space, time, or force, described in this book. In many respects this is the most attractive aspect of Einstein's work; it deduces a great number of remarkable phenomena solely from two general principles, aided by a mathematical calculus of great power; and it leaves aside as irrelevant all questions of mechanism. But this mode of development of the theory cannot be described in a non-technical book.

To avoid mathematical analysis we have had to resort to geometrical illustrations, which run parallel with the mathematical development and enable its processes to be understood to some extent. The question arises, are these merely illustrations of the mathematical argument, or illustrations of the actual processes of nature. No doubt the safest course is to avoid the thorny questions raised by the latter suggestion, and to say that it is quite sufficient that the illustrations should correctly replace the mathematical argument. But I think that this would give a misleading view of what the theory of relativity has accomplished in science.

The physicist, so long as he thinks as a physicist, has a definite belief in a real world outside him. For instance, he believes that atoms and molecules really exist; they are not mere inventions that enable him to grasp certain laws of chemical combination. That suggestion might have sufficed in the early days of the atomic theory; but now the existence of atoms as entities in the real world of physics is fully demonstrated. This confident assertion is not inconsistent with philosophic doubts as to the meaning of ultimate reality.

When therefore we are asked whether the four-dimensional world may not be regarded merely as an illustration of mathematical processes, we must bear in mind that our questioner has probably an ulterior motive. He has already a belief in a real world of three Euclidean dimensions, and he hopes to be allowed to continue in this belief undisturbed. In that case our answer must be definite; the real three-dimensional world is obsolete, and must be replaced by the four-dimensional space-time with non-Euclidean properties. In this book we have sometimes employed illustrations which certainly do not correspond to any physical reality—imaginary time, and an unperceived fifth dimension. But the four-dimensional world is no mere illustration; it is the real world of physics, arrived at in the recognised way by which physics has always (rightly or wrongly) sought for reality.

I hold a certain object before me, and see an outline of the figure of Britannia; another observer on the other side sees a picture of a monarch; a third observer sees only a thin rectangle. Am I to say that the figure of Britannia is the real object; and that the crude impressions of the other observers must be corrected to make allowance for their positions? All the appearances can be accounted for if we are all looking at a three-dimensional object—a penny—and no reasonable person can doubt that the penny is the corresponding physical reality. Similarly, an observer on the earth sees and measures an oblong block; an observer on another star contemplating the same object finds it to be a cube. Shall we say that the oblong block is the real thing, and that the other observer must correct his measures to make allowance for his motion? All the appearances are accounted for if the real object is four-dimensional, and the observers are merely measuring different three-dimensional appearances or sections; and it seems impossible to doubt that this is the true explanation. He who doubts the reality of the four-dimensional world (for logical, as distinct from experimental, reasons) can only be compared to a man who doubts the reality of the penny, and prefers to regard one of its innumerable appearances as the real object.

Physical reality is the synthesis of all possible physical aspects of nature. An illustration may be taken from the phenomena of radiant-energy, or light. In a very large number of phenomena the light coming from an atom appears to be a series of spreading waves, extending so as to be capable of filling the largest telescope yet made. In many other phenomena the light coming from an atom appears to remain a minute bundle of energy, all of which can enter and blow up a single atom. There may be some illusion in these experimental deductions; but if not, it must be admitted that the physical reality corresponding to light must be some synthesis comprehending both these appearances. How to make this synthesis has hitherto baffled conception. But the lesson is that a vast number of appearances may be combined into one consistent whole—perhaps all appearances that are directly perceived by terrestrial observers—and yet the result may still be only an appearance. Reality is only obtained when all conceivable points of view have been combined.

That is why it has been necessary to give up the reality of the everyday world of three dimensions. Until recently it comprised all the possible appearances that had been considered. But now it has been discovered that there are new points of view with new appearances; and the reality must contain them all. It is by bringing in all these new points of view that we have been able to learn the nature of the real world of physics.

Let us briefly recapitulate the steps of our synthesis. We found one step already accomplished. The immediate perception of the world with one eye is a two-dimensional appearance. But we have two eyes, and these combine the appearances of the world as seen from two positions; in some mysterious way the brain makes the synthesis by suggesting solid relief, and we obtain the familiar appearance of a three-dimensional world. This suffices for all possible positions of the observer within the parts of space hitherto explored. The next step was to combine the appearances for all possible states of uniform motion of the observer. The result was to add another dimension to the world, making it four-dimensional. Next the synthesis was extended to include all possible variable motions of the observer. The process of adding dimensions stopped, but the world became non-Euclidean; a new geometry called Riemannian geometry was adopted. Finally the points of view of observers varying in size in any way were added; and the result was to replace the Riemannian geometry by a still more general geometry described in the last chapter.

The search for physical reality is not necessarily utilitarian, but it has been by no means profitless. As the geometry became more complex, the physics became simpler; until finally it almost appears that the physics has been absorbed into the geometry. We did not consciously set out to construct a geometrical theory of the world; we were seeking physical reality by approved methods, and this is what has happened.

Is the point now reached the ultimate goal? Have the points of view of all conceivable observers now been absorbed? We do not assert that they have. But it seems as though a definite task has been rounded off, and a natural halting-place reached. So far as we know, the different possible impersonal points of view have been exhausted—those for which the observer can be regarded as a mechanical automaton, and can be replaced by scientific measuring-appliances. A variety of more personal points of view may indeed be needed for an ultimate reality; but they can scarcely be incorporated in a real world of physics. There is thus justification for stopping at this point but not for stopping earlier.

It may be asked whether it is necessary to take into account all conceivable observers, many of whom, we suspect, have no existence. Is not the *real* world that which comprehends the appearances to all *real* observers? Whether or not it is a tenable hypothesis that that which no one observes does not exist, science uncompromisingly rejects it. If we deny the rights of extra-terrestrial observers, we must take the side of the Inquisition against Galileo. And if extra-terrestrial observers are admitted, the other observers, whose results are here combined, cannot be excluded.

Our inquiry into the nature of things is subject to certain limitations which it is important to realise. The best comparison I can offer is with a future antiquarian investigation, which may be dated about the year 5000 A.D. An interesting find has been made relating to a vanished civilisation which flourished about the twentieth century, namely a volume containing a large number of games of chess, written out in the obscure symbolism usually adopted for that purpose. The antiquarians, to whom the game was hitherto unknown, manage to discover certain uniformities; and by long research they at last succeed in establishing beyond doubt the nature of the moves and rules of the game. But it is obvious that no amount of study of the volume will reveal the true nature either of the participants in the game—the chessmen—or the field of the game—the chess-board. With regard to the former, all that is possible is to give arbitrary names distinguishing the chessmen according to their properties; but with regard to the chess-board something more can be stated. The material of the board is unknown, so too are the shapes of the meshes—whether squares or diamonds; but it is ascertainable that the different points of the field are connected with one another by relations of two-dimensional order, and a large number of hypothetical types of chess-board satisfying these relations of order can be constructed. In spite of these gaps in their knowledge, our antiquarians may fairly claim that they thoroughly understand the game of chess.

The application of this analogy is as follows. The recorded games are our physical experiments. The rules of the game, ascertained by study of them, are the laws of physics. The hypothetical chess-board of 64 squares is the space and time of some particular observer or player; whilst the more general relations of two-fold order, are the absolute relations of order in space-time which we have been studying. The chessmen are the entities of physics—electrons, particles, or point-events; and the range of movement may perhaps be compared to the fields of relation radiating from them—electric and gravitational fields, or intervals. By no amount of study of the experiments can the absolute nature or appearance of these participants be deduced; nor is this knowledge relevant, for without it we may yet learn "the game" in all its intricacy. Our knowledge of the nature of things must be like the antiquarians' knowledge of the nature of chessmen, viz. their nature as pawns and pieces in the game, not as carved shapes of wood. In the latter aspect they may have relations and significance transcending anything dreamt of in physics.

It is believed that the familiar things of experience are very complex; and the scientific method is to analyse them into simpler elements. Theories and laws of behaviour of these simpler constituents are studied; and from these it becomes possible to predict and explain phenomena. It seems a natural procedure to explain the complex in terms of the simple, but it carries with it the necessity of explaining the familiar in terms of the unfamiliar.

There are thus two reasons why the ultimate constituents of the real world must be of an unfamiliar nature. Firstly, all familiar objects are of a much too complex character. Secondly, familiar objects belong not to the real world of physics, but to a much earlier stage in the synthesis of appearances. The ultimate elements in a theory of the world must be of a nature impossible to define in terms recognisable to the mind.

The fact that he has to deal with entities of unknown nature presents no difficulty to the mathematician. As the mathematician in the Prologue explained, he is never so happy as when he does not know what he is talking about. But we ourselves cannot take any interest in the chain of reasoning he is producing, unless we can give it some meaning—a meaning, which we find by experiment, it will bear. We have to be in a position to make a sort of running comment on his work. At first his symbols bring no picture of anything before our eyes, and we watch in silence. Presently we can say "Now he is talking about a particle of matter"... "Now he is talking about another particle"... "Now he is saying where they will be at a certain time of day"... "Now he says that they will be in the same spot at a certain time." We watch to see.—"Yes. The two particles have collided. For once he is speaking about something familiar, and speaking the truth, although, of course, he does not know it." Evidently his chain of symbols can be interpreted as describing what occurs in the world; we need not, and do not, form any idea of the meaning of each individual symbol; it is only certain elaborate combinations of them that we recognise.

Thus, although the elementary concepts of the theory are of undefined nature, at some later stage we must link the derivative concepts to the familiar objects of experience.

We shall now collect the results arrived at in the previous chapters by successive steps, and set the theory out in more logical order. The extension in Chapter 11 will not be considered here, partly because it would increase the difficulty of grasping the main ideas, partly because it is less certainly established.

In the relativity theory of nature the most elementary concept is the *point-event*. In ordinary language a point-event is an instant of time at a point of space; but this is only one aspect of the point-event, and it must not be taken as a definition. Time and space—the familiar terms—are derived concepts to be introduced much later in our theory. The first simple concepts are necessarily undefinable, and their nature is beyond human understanding. The aggregate of all the point-events is called the *world*. It is postulated that the world is four-dimensional, which means that a particular point-event has to be specified by the values of four variables or coordinates, though there is entire freedom as to the way in which these four identifying numbers are to be assigned.

The meaning of the statement that the world is four-dimensional is not so clear as it appears at first. An aggregate of a large number of things has in itself no particular number of dimensions. Consider, for example, the words on this page. To a casual glance they form a two-dimensional distribution; but they were written in the hope that the reader would regard them as a one-dimensional distribution. In order to define the number of dimensions we have to postulate some ordering relation; and the result depends entirely on what this ordering relation is—whether the words are ordered according to sense or to position on the page. Thus the statement that the world is four-dimensional contains an implicit reference to some ordering relation. This relation appears to be the *interval*, though I am not sure whether that alone suffices without some relation corresponding to *proximity*. It must be remembered that if the interval s between two events is small, the events are not necessarily near together in the ordinary sense.

Between any two neighbouring point-events there exists a certain relation known as the *interval* between them. The relation is a quantitative one which can be measured on a definite scale of numerical values[1]. But the term "interval" is not to be taken as a guide to the real nature of the relation, which is altogether beyond our conception. Its geometrical properties, which we have dwelt on so often in the previous chapters, can only represent one aspect of the relation. It may have other aspects associated with features of the world outside the scope of physics. But in physics we are concerned not with the nature of the relation but

[1]There is also a qualitative distinction into two kinds, ultimately identified as time-like and space-like, which for mathematical treatment are distinguished by real and imaginary numbers.

with the number assigned to express its intensity; and this suggests a graphical representation, leading to a geometrical theory of the world of physics.

What we have here called the *world* might perhaps have been legitimately called the *aether*; at least it is the universal substratum of things which the relativity theory gives us in place of the aether.

We have seen that the number expressing the intensity of the interval-relation can be measured practically with scales and clocks. Now, I think it is improbable that our coarse measures can really get hold of the individual intervals of point-events; our measures are not sufficiently microscopic for that. The interval which has appeared in our analysis must be a *macroscopic* value; and the potentials and kinds of space deduced from it are averaged properties of regions, perhaps small in comparison even with the electron, but containing vast numbers of the primitive intervals. We shall therefore pass at once to the consideration of the macroscopic interval; but we shall not forestall later results by assuming that it is measurable with a scale and clock. That property must be introduced in its logical order.

Consider a small portion of the world. It consists of a large (possibly infinite) number of point-events between every two of which an interval exists. If we are given the intervals between a point A and a sufficient number of other points, and also between B and the same points, can we calculate what will be the interval between A and B? In ordinary geometry this would be possible; but, since in the present case we know nothing of the relation signified by the word interval, it is impossible to predict any law *a priori*. But we have found in our previous work that there is such a rule, expressed by the formula

$$ds^2 = g_{11}\,dx_1^2 + g_{22}\,dx_2^2 + \cdots + 2g_{12}\,dx_1dx_2 + \cdots .$$

This means that, having assigned our identification numbers (x_1, x_2, x_3, x_4) to the point-events, we have only to measure ten different intervals to enable us to determine the ten coefficients, g_{11}, etc., which in a small region may be considered to be constants; then all other intervals in this region can be predicted from the formula. For any other region we must make fresh measures, and determine the coefficients for a new formula.

I think it is unlikely that the *individual* interval-relations of point-events follow any such definite rule. A microscopic examination would probably show them as quite arbitrary, the relations of so-called intermediate points being not necessarily intermediate. Perhaps even the primitive interval is not quantitative, but simply 1 for certain pairs of point-events and 0 for others. The formula given is just an average summary which suffices for our coarse methods of investigation, and holds true only statistically. Just as statistical averages of one community may differ from those of another, so may this statistical formula for one region of the world differ from that of another. This is the starting point of the infinite variety of nature.

Perhaps an example may make this clearer. Compare the point-events to persons, and the intervals to the degree of acquaintance between them. There is no means of forecasting the degree of acquaintance between A and B from a knowledge of the familiarity of both with C, D, E, etc. But a statistician may compute in any community a kind of average rule. In most cases if A and B

both know C, it slightly increases the probability of their knowing one another. A community in which this correlation was very high would be described as *cliquish*. There may be differences among communities in this respect, corresponding to their degree of cliquishness; and so the statistical laws may be the means of expressing intrinsic differences in communities.

Now comes the difficulty which is by this time familiar to us. The ten g's are concerned, not only with intrinsic properties of the world, but with our arbitrary system of identification-numbers for the point-events; or, as we have previously expressed it, they describe not only the kind of space-time, but the nature of the arbitrary mesh-system that is used. Mathematics shows the way of steering through this difficulty by fixing attention on expressions called tensors, of which $B^{\rho}_{\mu\nu\sigma}$ and $G_{\mu\nu}$ are examples.

A tensor does not express explicitly the measure of an intrinsic quality of the world, for some kind of mesh-system is essential to the idea of measurement of a property, except in certain very special cases where the property is expressed by a single number termed an invariant, e.g. the interval, or the total curvature. But to state that a tensor vanishes, or that it is equal to another tensor in the same region, is a statement of intrinsic property, quite independent of the mesh-system chosen. Thus by keeping entirely to tensors, we contrive that there shall be behind our formulae an undercurrent of information having reference to the intrinsic state of the world.

In this way we have found two absolute formulae, which appear to be fully confirmed by observation, namely
in empty space,

$$G_{\mu\nu} = 0,$$

in space containing matter,

$$G_{\mu\nu} = K_{\mu\nu},$$

where $K_{\mu\nu}$ contains only physical quantities which are perfectly familiar to us, viz. the density and state of motion of the matter in the region.

I think the usual view of these equations would be that the first expresses some law existing in the world, so that the point-events by natural necessity tend to arrange their relations in conformity with this equation. But when matter intrudes it causes a disturbance or strain of the natural linkages; and a rearrangement takes place to the extent indicated by the second equation.

But let us examine more closely what the equation $G_{\mu\nu} = 0$ tells us. We have been giving the mathematician a free hand with his indefinable intervals and point-events. He has arrived at the quantity $G_{\mu\nu}$; but as yet this means to us—absolutely nothing. The pure mathematician left to himself never "deviates into sense." His work can never relate to the familiar things around us, unless we boldly lay hold of some of his symbols and *give* them an intelligible meaning—tentatively at first, and then definitely as we find that they satisfy all experimental knowledge. We have decided that in empty space $G_{\mu\nu}$ vanishes. Here is our opportunity. In default of any other suggestion as to what the vanishing of $G_{\mu\nu}$ might mean, let us say that the vanishing of $G_{\mu\nu}$ *means* emptiness; so that $G_{\mu\nu}$, if it does not vanish, is a condition of the world which distinguishes space said to be occupied from space said to be empty. Hitherto $G_{\mu\nu}$ was merely a formal outline to be filled with some undefined contents; we are as far as ever from being able to explain what those contents are; but we have now given a recognisable meaning to the completed picture, so that we shall know it when we come across it in the familiar world of experience.

The two equations are accordingly merely definitions—definitions of the way in which certain states of the world (described in terms of the indefinables) impress themselves on our perceptions. When we perceive that a certain region of the world is empty, that is merely the mode in which our senses recognise that it is curved no higher than the first degree. When we perceive that a region contains matter we are recognising the intrinsic curvature of the world; and when we believe we are measuring the mass and momentum of the matter (relative to some axes of reference) we are measuring certain components of world-curvature (referred to those axes). The statistical averages of something unknown, which have been used to describe the state of the world, vary from point to point; and it is out of these that the mind has constructed the familiar notions of matter and emptiness.

The law of gravitation is not a law in the sense that it restricts the possible behaviour of the substratum of the world; it is merely the definition of a vacuum. We need not regard matter as a foreign entity causing a disturbance in the gravitational field; the disturbance is matter. In the same way we do not regard light as an intruder in the electromagnetic field, causing the electromagnetic force to oscillate along its path; the oscillations constitute the light. Nor is heat a fluid causing agitation of the molecules of a body; the agitation is heat.

This view, that matter is a symptom and not a cause, seems so natural that it is surprising that it should be obscured in the usual presentation of the theory. The reason is that the connection of mathematical analysis with the things of experience is usually made, not by determining what matter is, but by what certain combinations of matter do. Hence the interval is at once identified with something familiar to experience, namely the thing that a scale and a clock measure. However advantageous that may be for the sake of bringing the theory into touch with experiment at the outset, we can scarcely hope to build up a theory of the nature of things if we take a scale and clock as the simplest unanalysable concepts. The result of this logical inversion is that by the time the equation $G_{\mu\nu} = K_{\mu\nu}$ is encountered, both sides of the equation are well-defined quantities. Their *necessary* identity is overlooked, and the equation is regarded as a new law of nature. This is the fault of introducing the scale and clock prematurely. For our part we prefer first to define what matter is in terms of the elementary concepts of the theory; then we can introduce any kind of scientific apparatus; and finally determine what property of the world that apparatus will measure.

Matter defined in this way obeys all the laws of mechanics, including conservation of energy and momentum. Proceeding with a similar development of Weyl's more general theory of the combined gravitational and electrical fields, we should find that it has the familiar electrical and optical properties. It is purely gratuitous to suppose that there is anything else present, controlling but not to be identified with the relations of the fourteen potentials (g's and κ's).

There is only one further requirement that can be demanded from matter.

Our brains are constituted of matter, and they feel and think—or at least feeling and thinking are closely associated with motions or changes of the matter of the brain. It would be difficult to say that any hypothesis as to the nature of matter makes this process less or more easily understood; and a brain constituted out of differential coefficients of g's can scarcely be said to be less adapted to the purposes of thought than one made, say, out of tiny billiard balls! But I think we may even go a little beyond this negative justification. The primary interval relation is of an undefined nature, and the g's contain this undefinable element. The expression $G_{\mu\nu}$ is therefore of defined *form*, but of undefined *content*. By its form alone it is fitted to account for all the physical properties of matter; and physical investigation can never penetrate beneath the form. The matter of the brain in its physical aspects is merely the form; but the reality of the brain includes the content. We cannot expect the form to explain the activities of the content, any more than we can expect the number 4 to explain the activities of the Big Four at Versailles.

Some of these views of matter were anticipated with marvellous foresight by W. K. Clifford forty years ago. Whilst other English physicists were distracted by vortex-atoms and other will-o'-the-wisps, Clifford was convinced that matter and the motion of matter were aspects of space-curvature *and nothing more.* And he was no less convinced that these geometrical notions were only partial aspects of the relations of what he calls "elements of feeling."—"The reality corresponding to our perception of the motion of matter is an element of the complex thing we call feeling. What we might perceive as a plexus of nerve-disturbances is really in itself a feeling; and the succession of feelings which constitutes a man's consciousness is the reality which produces in our minds the perception of the motions of his brain. These elements of feeling have relations of *nextness* or contiguity in space, which are exemplified by the sight-perceptions of contiguous points; and relations of succession in time which are exemplified by all perceptions. Out of these two relations the future theorist has to build up the world as best he may. Two things may perhaps help him. There are many lines of mathematical thought which indicate that distance or quantity may come to be expressed in terms of *position* in the wide sense of the *analysis situs.* And the theory of space-curvature hints at a possibility of describing matter and motion in terms of extension only." (*Fortnightly Review*, 1875.)

The equation $G_{\mu\nu} = K_{\mu\nu}$ is a kind of dictionary explaining what the different components of world-curvature mean in terms ordinarily used in mechanics. If we write it in the slightly modified, but equivalent, form

$$G_{\mu\nu} - \tfrac{1}{2} g_{\mu\nu} G = -8\pi T_{\mu\nu},$$

we have the following scheme of interpretation

$$\begin{array}{cccc} T_{11}, & T_{12}, & T_{13}, & T_{14} \\ & T_{22}, & T_{23}, & T_{24} \\ & & T_{33}, & T_{34} \\ & & & T_{44} \end{array} = \begin{array}{cccc} p_{11} + \rho u^2, & p_{12} + \rho uv, & p_{13} + \rho uw, & -\rho u, \\ & p_{22} + \rho v^2, & p_{23} + \rho vw, & -\rho v, \\ & & p_{33} + \rho w^2, & -\rho w, \\ & & & \rho. \end{array}$$

Here we are using the partitions of space and time adopted in ordinary mechanics; ρ is the density of the matter, u, v, w its component velocities, and p_{11},

$p_{12}, \ldots p_{33}$, the components of the internal stresses which are believed to be analysable into molecular movements.

Now the question arises, is it legitimate to make identifications on such a wholesale scale? Having identified T_{44} as density, can we go on to identify another quantity T_{34} as density multiplied by velocity? It is as though we identified one "thing" as *air*, and a quite different "thing" as *wind*. Yes, it is legitimate, because we have not hitherto explained what is to be the counterpart of velocity in our scheme of the world; and this is the way we choose to introduce it. All identifications are at this stage provisional, being subject to subsequent test by observation.

A definition of the velocity of matter in some such terms as "*wind* divided by *air*," does not correspond to the way in which motion primarily manifests itself in our experience. Motion is generally recognised by the disappearance of a particle at one point of space and the appearance of an apparently identical particle at a neighbouring point. This manifestation of motion can be deduced mathematically from the identifying definition here adopted. Remembering that in physical theory it is necessary to proceed from the simple to the complex, which is often opposed to the instinctive desire to proceed from the familiar to the unfamiliar, this inversion of the order in which the manifestations of motion appear need occasion no surprise. Permanent identity of particles of matter (without which the ordinary notion of velocity fails) is a very familiar idea, but it appears to be a very complex feature of the world.

A simple instance may be given where the familiar kinematical conception of motion is insufficient. Suppose a perfectly homogeneous continuous ring is rotating like a wheel, what meaning can we attach to its motion? The kinematical conception of motion implies change—disappearance at one point and reappearance at another point—but no change is detectable. The state at any one moment is the same as at a previous moment, and the matter occupying one position now is indistinguishable from the matter in the same position a moment ago. At the most it can only differ in a mysterious non-physical quality—that of identity; but if, as most physicists are willing to believe, matter is some state in the aether, what can we mean by saying that two states are exactly alike, but are not identical? Is the hotness of the room equal to, but not identical with, its hotness yesterday? Considered kinematically, the rotation of the ring appears to have no meaning; yet the revolving ring differs mechanically from a stationary ring. For example, it has gyrostatic properties. The fact that in nature a ring has atomic and not continuous structure is scarcely relevant. A conception of motion which affords a distinction between a rotating and non-rotating continuous ring must be possible; otherwise this would amount to an *a priori* proof that matter is atomic. According to the conception now proposed, velocity of matter is as much a static quality as density. Generally velocity is accompanied by changes in the physical state of the world, which afford the usual means of recognising its existence; but the foregoing illustration shows that these symptoms do not always occur.

This definition of velocity enables us to understand why velocity except in reference to matter is meaningless, whereas acceleration and rotation have a meaning. The philosophical argument, that velocity through space is meaningless, ceases to apply as soon as we admit any kind of structure or aether in empty regions; consequently the problem is by no means so simple as is often supposed. But our definition of velocity is dynamical, not kinematical. Velocity is the ratio of certain components of $T_{\mu\nu}$, and only exists when T_{44} is not zero. Thus matter (or electromagnetic energy) is the only thing that can have a velocity relative to the frame of reference. The velocity of the world-structure or aether, where the $T_{\mu\nu}$ vanish, is always of the indeterminate form 00. On the other hand acceleration and rotation are defined by means of the $g_{\mu\nu}$ and exist wherever these exist[2]; so that the acceleration and rotation of the world-structure or aether relative to the frame of reference are determinate. Notice that acceleration is not defined as change of velocity; it is an independent entity, much simpler and more universal than velocity. It is from a comparison of these two entities that we ultimately obtain the definition of time.

This finally resolves the difficulty encountered in Chapter 10—the apparent difference in the Principle of Relativity as applied to uniform and non-uniform motion. Fundamentally velocity and acceleration are both static qualities of a region of the world (referred to some mesh-system). Acceleration is a comparatively simple quality present wherever there is geodesic structure, that is to say everywhere. Velocity is a highly complex quality existing only where the structure is itself more than ordinarily complicated, viz. in matter. Both these qualities commonly give physical manifestations, to which the terms acceleration and velocity are more particularly applied; but it is by examining their more fundamental meaning that we can understand the universality of the one and the localisation of the other.

It has been shown that there are four identical relations between the ten qualities of a piece of matter here identified, which depend solely on the way the $G_{\mu\nu}$ were by definition constructed out of simpler elements. These four relations state that, *provided the mesh-system is drawn in one of a certain number of ways*, mass (or energy) and momentum will be conserved. The conservation of mass is of great importance; matter will be permanent, and for every particle disappearing at any point a corresponding mass will appear at a neighbouring point; the change consists in the displacement of matter, not its creation or destruction. This gives matter the right to be regarded, not as a mere assemblage of symbols, but as the substance of a permanent world. But the permanent world so found demands the partitioning of space-time in one of a certain number of ways, viz. those discussed in Chapter 3;[3] from these a particular space and time are selected, because the observer wishes to consider himself, or some arbitrary body, at rest. This gives the space and time used for ordinary descriptions of experience. In this way we are able to introduce perceptual space and time into the four-dimensional world, as derived concepts depending on our desire that the new-found matter should be permanent.

---

[2]Even in Newtonian mechanics we speak of the "field of acceleration," and think of it as existing even when there is no test body to display the acceleration. In the present theory this field of acceleration is described by the $g_{\mu\nu}$. There is no such thing as a "field of velocity" in empty space; but there is in a material ocean.

[3]When the kind of space-time is such that a strict partition of this kind is impossible, strict conservation does not exist; but we retain the principle as formally satisfied by attributing energy and momentum to the gravitational field.

I think it is now possible to discern something of the reason why the world must of necessity be as we have described it. When the eye surveys the tossing waters of the ocean, the eddying particles of water leave little impression; it is the waves that strike the attention, because they have a certain degree of permanence. The motion particularly noticed is the motion of the wave-form, which is not a motion of the water at all. So the mind surveying the world of point-events looks for the permanent things. The simpler relations, the intervals and potentials, are transient, and are not the stuff out of which mind can build a habitation for itself. But the thing that has been identified with matter is permanent, and because of its permanence it must be for mind the substance of the world. Practically no other choice was possible.

It must be recognised that the conservation of mass is not exactly equivalent to the permanence of matter. If a loaf of bread suddenly transforms into a cabbage, our surprise is not diminished by the fact that there may have been no change of weight. It is not very easy to define this extra element of permanence required, because we accept as quite natural apparently similar transformations—an egg into an omelette, or radium into lead. But at least it seems clear that some degree of permanence of one quality, mass, would be the primary property looked for in matter, and this gives sufficient reason for the particular choice.

We see now that the choice of a permanent substance for the world of perception necessarily carries with it the law of gravitation, all the laws of mechanics, and the introduction of the ordinary space and time of experience. Our whole theory has really been a discussion of the most general way in which permanent substance can be built up out of relations; and it is the mind which, by insisting on regarding only the things that are permanent, has actually imposed these laws on an indifferent world. Nature has had very little to do with the matter; she had to provide a basis—point-events; but practically anything would do for that purpose if the relations were of a reasonable degree of complexity. The relativity theory of physics reduces everything to relations; that is to say, it is structure, not material, which counts. The structure cannot be built up without material; but the nature of the material is of no importance. We may quote a passage from Bertrand Russell's *Introduction to Mathematical Philosophy.*

"There has been a great deal of speculation in traditional philosophy which might have been avoided if the importance of structure, and the difficulty of getting behind it, had been realised. For example it is often said that space and time are subjective, but they have objective counterparts; or that phenomena are subjective, but are caused by things in themselves, which must have differences *inter se* corresponding with the differences in the phenomena to which they give rise. Where such hypotheses are made, it is generally supposed that we can know very little about the objective counterparts. In actual fact, however, if the hypotheses as stated were correct, the objective counterparts would form a world having the same structure as the phenomenal world.... In short, every proposition having a communicable significance must be true of both worlds or of neither: the only difference must lie in just that essence of individuality which always eludes words and baffles description, but which for that very reason is irrelevant to science."

This is how our theory now stands.—We have a world of point-events with their primary interval-relations. Out of these an unlimited number of more complicated relations and qualities can be built up mathematically, describing various features of the state of the world. These exist in nature in the same sense as an unlimited number of walks exist on an open moor. But the existence is, as it were, latent unless someone gives a significance to the walk by following it; and in the same way the existence of any one of these qualities of the world only acquires significance above its fellows, if a mind singles it out for recognition. Mind filters out matter from the meaningless jumble of qualities, as the prism filters out the colours of the rainbow from the chaotic pulsations of white light. Mind exalts the permanent and ignores the transitory; and it appears from the mathematical study of relations that the only way in which mind can achieve her object is by picking out one particular quality as the permanent substance of the perceptual world, partitioning a perceptual time and space for it to be permanent in, and, as a necessary consequence of this Hobson's choice, the laws of gravitation and mechanics and geometry have to be obeyed. Is it too much to say that mind's search for permanence has created the world of physics? So that the world we perceive around us could scarcely have been other than it is[4]?

The last sentence possibly goes too far, but it illustrates the direction in which these views are tending. With Weyl's more general theory of interval-relations, the laws of electrodynamics appear in like manner to depend merely on the identification of another permanent thing—electric charge. In this case the identification is due, not to the rudimentary instinct of the savage or the animal, but the more developed reasoning-power of the scientist. But the conclusion is that the whole of those laws of nature which have been woven into a unified scheme—mechanics, gravitation, electrodynamics and optics—have their origin, not in any special mechanism of nature, but in the workings of the mind.

"Give me matter and motion," said Descartes, "and I will construct the universe." The mind reverses this. "Give me a world—a world in which there are relations—and I will construct matter and motion."

Are there then no genuine laws in the external world? Laws inherent in the substratum of events, which break through into the phenomena otherwise regulated by the despotism of the mind? We cannot foretell what the final answer will be; but, at present, we have to admit that there are laws which appear to have their seat in external nature. The most important of these, if not the only law, is a law of atomicity. Why does that quality of the world which distinguishes matter from emptiness exist only in certain lumps called atoms or electrons, all of comparable mass? Whence arises this discontinuity? At present, there seems no ground for believing that discontinuity is a law due to the mind; indeed the mind seems rather to take pains to smooth the discontinuities of nature into continuous perception. We can only suppose that there is something in the nature of things that causes this aggregation into atoms. Probably our analysis into point-events is not final; and if it could be pushed further to reach

[4]This summary is intended to indicate the direction in which the views suggested by the relativity theory appear to me to be tending, rather than to be a precise statement of what has been established. I am aware that there are at present many gaps in the argument. Indeed the whole of this part of the discussion should be regarded as suggestive rather than dogmatic.

something still more fundamental, then atomicity and the remaining laws of physics would be seen as identities. This indeed is the only kind of explanation that a physicist could accept as ultimate. But this more ultimate analysis stands on a different plane from that by which the point-events were reached. The world *may* be so constituted that the laws of atomicity must necessarily hold; but, so far as the mind is concerned, there seems no reason why it should have been constituted in that way. We can conceive a world constituted otherwise. But our argument hitherto has been that, however the world is constituted, the necessary combinations of things can be found which obey the laws of mechanics, gravitation and electrodynamics, and these combinations are ready to play the part of the world of perception for any mind that is tuned to appreciate them; and further, any world of perception of a different character would be rejected by the mind as unsubstantial.

If atomicity depends on laws inherent in nature, it seems at first difficult to understand why it should relate to matter especially; since matter is not of any great account in the analytical scheme, and owes its importance to irrelevant considerations introduced by the mind. It has appeared, however, that atomicity is by no means confined to matter and electricity; the quantum, which plays so great a part in recent physics, is apparently an atom of action. So nature cannot be accused of connivance with mind in singling out matter for special distinction. Action is generally regarded as the most fundamental thing in the real world of physics, although the mind passes it over because of its lack of permanence; and it is vaguely believed that the atomicity of action is the general law, and the appearance of electrons is in some way dependent on this. But the precise formulation of the theory of quanta of action has hitherto baffled physicists.

There is a striking contrast between the triumph of the scientific mind in formulating the great general scheme of natural laws, nowadays summed up in the principle of least action, and its present defeat by the newly discovered but equally general phenomena depending on the laws of atomicity of quanta. It is too early to cry failure in the latter case; but possibly the contrast is significant. It is one thing for the human mind to extract from the phenomena of nature the laws which it has itself put into them; it may be a far harder thing to extract laws over which it has had no control. It is even possible that laws which have not their origin in the mind may be irrational, and we can never succeed in formulating them. This is, however, only a remote possibility; probably if they were really irrational it would not have been possible to make the limited progress that has been achieved. But if the laws of quanta do indeed differentiate the actual world from other worlds possible to the mind, we may expect the task of formulating them to be far harder than anything yet accomplished by physics.

The theory of relativity has passed in review the whole subject-matter of physics. It has unified the great laws, which by the precision of their formulation and the exactness of their application have won the proud place in human knowledge which physical science holds to-day. And yet, in regard to the nature of things, this knowledge is only an empty shell—a form of symbols. It is knowledge of structural form, and not knowledge of content. All through the physical world runs that unknown content, which must surely be the stuff of our consciousness. Here is a hint of aspects deep within the world of physics, and yet unattainable by the methods of physics. And, moreover, we have found that where science has progressed the farthest, the mind has but regained from nature that which the mind has put into nature.

We have found a strange foot-print on the shores of the unknown. We have devised profound theories, one after another, to account for its origin. At last, we have succeeded in reconstructing the creature that made the foot-print. And Lo! it is our own.

Appendix

Mathematical Notes

The references marked "Report" are to the writer's "Report on the Relativity Theory of Gravitation" for the Physical Society of London (Fleetway Press), where fuller mathematical details are given.

Probably the most complete treatise on the mathematical theory of the subject is H. Weyl's *Raum, Zeit, Materie* (Julius Springer, Berlin).

Note 1 (p. 15).

It is not possible to predict the contraction rigorously from the universally accepted electromagnetic equations, because these do not cover the whole ground. There must be other forces or conditions which govern the form and size of an electron; under electromagnetic forces alone it would expand indefinitely. The old electrodynamics is entirely vague as to these forces.

The theory of Larmor and Lorentz shows that if any system at rest in the aether is in equilibrium, a similar system in uniform motion through the aether, but with all lengths in the direction of motion diminished in FitzGerald's ratio, will also be in equilibrium so far as the differential equations of the electromagnetic field are concerned. There is thus a general theoretical agreement with the observed contraction, provided the boundary conditions at the surface of an electron behave in the same way. The latter suggestion is confirmed by experiments on isolated electrons in rapid motion (Kaufmann's experiment). It turns out that this requires an electron to suffer the same kind of contraction as a material rod; and thus, although the theory throws light on the adjustments involved in material contraction, it can scarcely be said to give an explanation of the occurrence of contraction generally.

Note 2 (p. 34).

Suppose a particle moves from (x_1, y_1, z_1, t_1) to (x_2, y_2, z_2, t_2), its velocity u is given by

$$u^2 = \frac{(x_2 - x_1)^2 + (y_2 - y_1)^2 + (z_2 - z_1)^2}{(t_2 - t_1)^2}.$$

Hence from the formula for s^2

$$s = (t_2 - t_1)\sqrt{(1 - u^2)}.$$

(We omit a $\sqrt{-1}$, as the sign of s^2 is changed later in the chapter.)

If we take t_1 and t_2 to be the start and finish of the aviator's cigar (Chapter 1), then as judged by a terrestrial observer, $t_2 - t_1 = 60$ minutes, $\sqrt{(1 - u^2)} =$ FitzGerald contraction $= \frac{1}{2}$.

As judged by the aviator,

$$t_2 - t_1 = 30 \text{ minutes}, \quad \sqrt{(1 - u^2)} = 1.$$

Thus for both observers $s = 30$ minutes, verifying that it is an absolute quantity independent of the observer.

Note 3 (p. 35).

The formulae of transformation to axes with a different orientation are

$$x = x' \cos\theta - \tau' \sin\theta, \quad y = y', \quad z = z', \quad \tau = x' \sin\theta + \tau' \cos\theta,$$

where θ is the angle turned through in the plane $x\tau$.

Let $u = i \tan\theta$, so that $\cos\theta = (1 - u^2)^{-\frac{1}{2}} = \beta$, say. The formulae become

$$x = \beta(x' - iu\tau'), \quad y = y', \quad z = z', \quad \tau = \beta(\tau' + iux'),$$

or, reverting to real time by setting $i\tau = t$,

$$x = \beta(x' - ut'), \quad y = y', \quad z = z', \quad t = \beta(t' - ux'),$$

which gives the relation between the estimates of space and time by two different observers.

The factor β gives in the first equation the FitzGerald contraction, and in the fourth equation the retardation of time. The terms ut' and ux' correspond to the changed conventions as to *rest* and *simultaneity*.

A point at rest, $x =$ const., for the first observer corresponds to a point moving with velocity u, $x' - ut' =$ const., for the second observer. Hence their relative velocity is u.

Note 4 (p. 60).

The condition for flat space in two dimensions is

$$\frac{\partial}{\partial x_1}\left(\frac{g_{12}}{g_{11}\sqrt{(g_{11}g_{22} - g_{12}^2)}}\frac{\partial g_{11}}{\partial x_2} - \frac{1}{\sqrt{(g_{11}g_{22} - g_{12}^2)}}\frac{\partial g_{22}}{\partial x_1}\right)$$
$$+ \frac{\partial}{\partial x_2}\left(\frac{2}{\sqrt{(g_{11}g_{22} - g_{12}^2)}}\frac{\partial g_{12}}{\partial x_1} - \frac{1}{\sqrt{(g_{11}g_{22} - g_{12}^2)}}\frac{\partial g_{11}}{\partial x_2}\right.$$
$$\left. - \frac{g_{12}}{g_{11}\sqrt{(g_{11}g_{22} - g_{12}^2)}}\frac{\partial g_{11}}{\partial x_1}\right) = 0.$$

Note 5 (p. 65).

Let g be the determinant of four rows and columns formed with the elements $g_{\mu\nu}$.

Let $g^{\mu\nu}$ be the minor of $g_{\mu\nu}$, divided by g.

Let the "3-index symbol" $\{\mu\nu, \lambda\}$ denote

$$\frac{1}{2}g^{\lambda\alpha}\left(\frac{\partial g_{\mu\alpha}}{\partial x_\nu} + \frac{\partial g_{\nu\alpha}}{\partial x_\mu} - \frac{\partial g_{\mu\nu}}{\partial x_\alpha}\right)$$

summed for values of α from 1 to 4. There will be 40 different 3-index symbols.

Then the Riemann-Christoffel tensor is

$$B^{\rho}_{\mu\nu\sigma} = \{\mu\sigma, \epsilon\}\{\epsilon\nu, \rho\} - \{\mu\nu, \epsilon\}\{\epsilon\sigma, \rho\} + \frac{\partial}{\partial x_\nu}\{\mu\sigma, \rho\} - \frac{\partial}{\partial x_\sigma}\{\mu\nu, \rho\},$$

the terms containing ϵ being summed for values of ϵ from 1 to 4.

The "contracted" Riemann-Christoffel tensor $G_{\mu\nu}$ can be reduced to

$$G_{\mu\nu} = -\frac{\partial}{\partial x_\alpha}\{\mu\nu, \alpha\} + \{\mu\alpha, \beta\}\{\nu\beta, \alpha\} + \frac{\partial^2}{\partial x_\mu \partial x_\nu}\log\sqrt{-g} - \{\mu\nu, \alpha\}\frac{\partial}{\partial x_\alpha}\log\sqrt{-g},$$

where in accordance with a general convention in this subject, each term containing a suffix twice over (α and β) must be summed for the values 1, 2, 3, 4 of that suffix.

The curvature $G = g^{\mu\nu}G_{\mu\nu}$, summed in accordance with the foregoing convention.

Note 6 (p. 70).

The electric potential due to a charge e is

$$\phi = \frac{e}{[r(1 - v_r/C)]},$$

where v_r is the velocity of the charge in the direction of r, C the velocity of light, and the square bracket signifies antedated values. To the first order of v_r/C, the denominator is equal to the *present* distance r, so the expression reduces to e/r in spite of the time of propagation. The foregoing formula for the potential was found by Liénard and Wiechert.

Note 7 (p. 71).

It is found that the following scheme of potentials rigorously satisfies the equations $G_{\mu\nu} = 0$, according to the values of $G_{\mu\nu}$ in Note 5,

$$\begin{matrix} -1/\gamma & 0 & 0 & 0 \\ & -{x_1}^2 & 0 & 0 \\ & & -{x_1}^2 \sin^2 {x_2}^2 & 0 \\ & & & \gamma \end{matrix}$$

where $\gamma = 1 - \kappa/x_1$ and κ is any constant (see Report, § 28). Hence these potentials describe a kind of space-time which can occur in nature referred to a possible mesh-system. If $\kappa = 0$, the potentials reduce to those for flat space-time referred to polar coordinates; and, since in the applications required κ will always be extremely small, our coordinates can scarcely be distinguished from polar coordinates. We can therefore use the familiar symbols r, θ, ϕ, t, instead of x_1, x_2, x_3, x_4. It must, however, be remembered that the identification with polar coordinates is only approximate; and, for example, an equally good approximation is obtained if we write $x_1 = r + \frac{1}{2}\kappa$, a substitution often used instead of $x_1 = r$ since it has the advantage of making the coordinate-velocity of light more symmetrical.

We next work out analytically all the mechanical and optical properties of this kind of space-time, and find that they agree observationally with those existing round a particle at rest at the origin with gravitational mass $\frac{1}{2}\kappa$. The conclusion is that the gravitational field here described is produced by a particle of mass $\frac{1}{2}\kappa$—or, if preferred, a particle of matter at rest is produced by the kind of space-time here described.

Note 8 (p. 73).

Setting the gravitational constant equal to unity, we have for a circular orbit

$$m/r^2 = v^2/r,$$

so that

$$m = v^2 r.$$

The earth's speed, v, is approximately 30 km. per sec., or $\frac{1}{10000}$ in terms of the velocity of light. The radius of its orbit, r, is about $1.5 10^8$ km. Hence, m, the gravitational mass of the sun is approximately 1.5 km.

The radius of the sun is 697,000 kms., so that the quantity $2m/r$ occurring in the formulae is, for the sun's surface, .00000424 or $0''.87$.

Note 9 (p. 91).

See Report, §§ 29, 30. The general equations of a geodesic are

$$\frac{d^2 x_\mu}{ds^2} + \{\alpha\beta, \mu\} \frac{dx_\alpha}{ds}\frac{dx_\beta}{ds} = 0 \quad (\mu = 1,\ 2,\ 3,\ 4).$$

From the formula for the line-element

$$ds^2 = -\gamma^{-1}\,dr^2 - r^2\,d\theta^2 + \gamma\,dt^2, \tag{1}$$

we calculate the three-index symbols and it is found that two of the equations of the geodesic take the rather simple form

$$\frac{d^2\theta}{ds^2} + \frac{2}{r}\frac{dr}{ds}\frac{d\theta}{ds} = 0,$$
$$\frac{d^2 t}{ds^2} + \frac{d(\log\gamma)}{dr}\frac{dr}{ds}\frac{dt}{ds} = 0,$$

which can be integrated giving

$$r^2\frac{d\theta}{ds} = h, \tag{2}$$
$$\frac{dt}{ds} = \frac{c}{\gamma}, \tag{3}$$

where h and c are constants of integration.

Eliminating dt and ds from (1), (2) and (3), we have

$$\left(\frac{h}{r^2}\frac{dr}{d\theta}\right)^2 + \frac{h^2}{r^2} = c^2 - 1 + \frac{2m}{r} + \frac{2mh^2}{r^3},$$

or writing $u = 1/r$,

$$\left(\frac{du}{d\theta}\right)^2 + u^2 = \frac{c^2-1}{h^2} + \frac{2mu}{h^2} + 2mu^3.$$

Differentiating with respect to θ

$$\frac{d^2u}{d\theta^2} + u = \frac{m}{h^2} + 3mu^2,$$

which gives the equation of the orbit in the usual form in particle dynamics. It differs from the equation of the Newtonian orbit by the small term $3mu^2$, which is easily shown to give the motion of perihelion.

The track of a ray of light is also obtained from this formula, since by the principle of equivalence it agrees with that of a material particle moving with the speed of light. This case is given by $ds = 0$, and therefore $h = \infty$. The differential equation for the path of a light-ray is thus

$$\frac{d^2u}{d\theta^2} + u = 3mu^2.$$

An approximate solution is

$$u = \frac{\cos\theta}{R} + \frac{m}{R^2}(\cos^2\theta + 2\sin^2\theta),$$

neglecting the very small quantity m^2/R^2. Converting to Cartesian coordinates, this becomes

$$x = R - \frac{m}{R}\frac{x^2+2y^2}{\sqrt{(x^2+y^2)}}.$$

The asymptotes of the light-track are found by taking y very large compared with x, giving

$$x = R\frac{2m}{R}y$$

so that the angle between them is $4m/R$.

Note 10 (p. 93).

Writing the line element in the form

$$ds^2 = -\left(1 + a\frac{m}{r} + \cdots\right)dr^2 - r^2\,d\theta^2 + \left(1 + b\frac{m}{r} + c\frac{m^2}{r^2} + \cdots\right)dt^2,$$

the approximate Newtonian attraction fixes b equal to -2; then the observed deflection of light fixes a equal to $+2$; and with these values the observed motion of Mercury fixes c equal to 0.

To insert an arbitrary coefficient of $r^2\,d\theta^2$ would merely vary the coordinate system. We cannot arrive at any intrinsically different kind of space-time in that way. Hence, within the limits of accuracy mentioned, the expression found by Einstein is completely determinable by observation.

It may be mentioned that the line-element

$$ds^2 = -dr^2 - r^2\,d\theta^2 + (1 - 2m/r)\,dt^2,$$

gives one-half the observed deflection of light, and one-third the motion of perihelion of Mercury. As both these can be obtained on older theories, taking account of the variation of mass with velocity, the coefficient γ^{-1} of dr^2 is the essentially novel point in Einstein's theory.

Note 11 (p. 97).

It is often supposed that by the Principle of Equivalence any invariant property which holds outside a gravitational field also holds in a gravitational field; but there is necessarily some limitation on this equivalence. Consider for instance the two invariant equations

$$ds^2 = 1,$$
$$ds^2\,(1 + k^4 B^{\rho}_{\mu\nu\sigma}\,B^{\mu\nu\sigma}_{\rho}) = 1,$$

where k is some constant having the dimensions of a length. Since $B^{\rho}_{\mu\nu\sigma}$ vanishes outside a gravitational field, if one of these equations is true the other will be. But they cannot both hold in a gravitational field, since there $B^{\rho}_{\mu\nu\sigma}B^{\mu\nu\sigma}_{\rho}$ does not vanish, and is in fact equal to $24m^2/r^6$. (I believe that the numerical factor 24 is correct; but there are 65,536 terms in the expression, and the terms which do not vanish have to be picked out.)

This ambiguity of the Principle of Equivalence is referred to in Report, §§ 14, 27; and an enunciation is given which makes it definite. The enunciation however is merely an explicit statement, and not a defence, of the assumptions commonly made in applying the principle.

So far as general reasoning goes there seems no ground for choosing ds^2 rather than $ds^2\,(1 + 24k^4m^2/r^6)$, or any similar expression, as the constant character in the vibration of an atom.

Note 12 (p. 98).

Let two rays diverging from a point at a distance R pass at distances r and $r + dr$ from a star of mass m. The deflection being $4m/r$, their divergence will be increased by $4m\,dr/r^2$. This increase will be equal to the original divergence dr/R if $r = \sqrt{4mR}$. Take for instance $4m = 10$ km., $R = 10^{15}$ km., then $r = 10^8$ km. So that the divergence of the light will be doubled, when the actual deflection of the ray is only 10^{-7}, or $0''.02$. In the case of a star seen behind the sun the added divergence has no time to take effect; but when the light has to travel a stellar distance after the divergence is produced, it becomes weakened by it. Generally in stellar phenomena the weakening of the light should be more prominent than the actual deflection.

Note 13 (p. 104).

The relations are (Report, § 39)

$$G^{\nu}_{\mu\nu} = \frac{1}{2}\frac{\partial G}{\partial x_{\mu}} \quad (\mu = 1,\ 2,\ 3,\ 4),$$

where $G^{\nu}_{\mu\nu}$ is the (contracted) covariant derivative of G^{ν}_{μ}, or $g^{\nu\alpha}G_{\mu\alpha}$.

I doubt whether anyone has performed the laborious task of verifying these identities by straightforward algebra.

Note 14 (p. 117).

The modified law for spherical space-time is in empty space

$$G_{\mu\nu} = \lambda g_{\mu\nu}.$$

In cylindrical space-time, matter is essential. The law in space occupied by matter is

$$G_{\mu\nu} - \frac{1}{2}g_{\mu\nu}(G - 2\lambda) = -8\pi T_{\mu\nu},$$

the term 2λ being the only modification. Spherical space-time of radius R is given by $\lambda = 3/R^2$; cylindrical space-time by $\lambda = 1/R^2$ provided matter of average density $\rho = 1/4\pi R^2$ is present. (See Report, §§ 50, 51.) The total mass of matter in the cylindrical world is $\frac{1}{2}\pi R$. This must be enormous, seeing that the sun's mass is only $1\frac{1}{2}$ kilometres.

Note 15 (p. 128).

Weyl's theory is given in *Berlin. Sitzungsberichte*, 30 May, 1918; *Annalen der Physik*, Bd. 59 (1919), p. 101.

Note 16 (p. 130).

The argument is rather more complicated than appears in the text, where the distinction between action-density and action in a region, curvature and total curvature in a region, has not been elaborated. Taking a definitely marked out region in space and time, its measured volume will be increased 16-fold by halving the gauge. Therefore for action-density we must take an expression which will be diminished 16-fold by halving the gauge. Now G is proportional to $1/R^2$, where R is the radius of curvature, and so is diminished 4-fold. The invariant

$B^{\rho}_{\mu\nu\sigma}B^{\mu\nu\sigma}_{\rho}$ has the same gauge-dimensions as G^2; and hence when integrated through a volume gives a pure number independent of the gauge. In Weyl's theory this is only the gravitational part of the complete invariant

$$(B^{\rho}_{\mu\nu\sigma} - \tfrac{1}{2}g^{\rho}_{\mu}F_{\nu\sigma})(B^{\mu\nu\sigma}_{\rho} - \tfrac{1}{2}g^{\mu}_{\rho}F^{\nu\sigma}),$$

which reduces to

$$B^{\rho}_{\mu\nu\sigma}B^{\mu\nu\sigma}_{\rho} + F_{\nu\sigma}F^{\nu\sigma}.$$

The second term gives actually the well-known expression for the action-density of the electromagnetic field, and this evidently strengthens the identification of this invariant with action-density.

Einstein's theory, on the other hand, creates a difficulty here, because although there may be action in an electromagnetic field without electrons, the curvature is zero.

Historical Note

Before the Michelson-Morley experiment the question had been widely discussed whether the aether in and near the earth was carried along by the earth in its motion, or whether it slipped through the interstices between the atoms. Astronomical aberration pointed decidedly to a stagnant aether; but the experiments of Arago and Fizeau on the effect of motion of transparent media on the velocity of light in those media, suggested a partial convection of the aether in such cases. These experiments were first-order experiments, i.e. they depended on the ratio of the velocity of the transparent body to the velocity of light. The Michelson-Morley experiment is the first example of an experiment delicate enough to detect second-order effects, depending on the square of the above ratio; the result, that no current of aether past terrestrial objects could be detected, appeared favourable to the view that the aether must be convected by the earth. The difficulty of reconciling this with astronomical aberration was recognised.

An attempt was made by Stokes to reconcile mathematically a convection of aether by the earth with the accurately verified facts of astronomical aberration; but his theory cannot be regarded as tenable. Lodge investigated experimentally the question whether smaller bodies carried the aether with them in their motion, and showed that the aether between two whirling steel discs was undisturbed.

The controversy, stagnant *versus* convected aether, had now reached an intensely interesting stage. In 1895, Lorentz discussed the problem from the point of view of the electrical theory of light and matter. By his famous transformation of the electromagnetic equations, he cleared up the difficulties associated with the first-order effects, showing that they could all be reconciled with a stagnant aether. In 1900, Larmor carried the theory as far as second-order effects, and obtained an exact theoretical foundation for FitzGerald's hypothesis of contraction, which had been suggested in 1892 as an explanation of the Michelson-Morley experiment. The theory of a stagnant aether was thus reconciled with all observational results; and henceforward it held the field.

Further second-order experiments were performed by Rayleigh and Brace on double refraction (1902, 1904), Trouton and Noble on a torsional effect on a charged condenser (1903), and Trouton and Rankine on electric conductivity

(1908). All showed that the earth's motion has no effect on the phenomena. On the theoretical side, Lorentz (1902) showed that the indifference of the equations of the electromagnetic field to any velocity of the axes of reference, which he had previously established to the first order, and Larmor to the second order, was exact to all orders. He was not, however, able to establish with the same exactness a corresponding transformation for bodies containing electrons.

Both Larmor and Lorentz had introduced a "local time" for the moving system. It was clear that for many phenomena this local time would replace the "real" time; but it was not suggested that the observer in the moving system would be deceived into thinking that it was the real time. Einstein, in 1905 founded the modern principle of relativity by postulating that this local time was *the time* for the moving observer; no real or absolute time existed, but only the local times, different for different observers. He showed that absolute simultaneity and absolute location in space are inextricably bound together, and the denial of the latter carries with it the denial of the former. By realising that an observer in the moving system would measure all velocities in terms of the local space and time of that system, Einstein removed the last discrepancies from Lorentz's transformation.

The relation between the space and time coordinates in two systems in relative motion was now obtained immediately from the principles of space and time-measurement. It must hold for all phenomena provided they do not postulate a medium which can serve as a standard for absolute location and simultaneity. The previous deduction of these formulae by lengthy transformation of the electromagnetic equations now appears as a particular case; it shows that electromagnetic phenomena have no reference to a medium with such properties.

The combination of the local spaces and times of Einstein into an absolute space-time of four dimensions is the work of Minkowski (1908). Chapter 3 is largely based on his researches. Much progress was made in the four-dimensional vector-analysis of the world; but the whole problem was greatly simplified when Einstein and Grossmann introduced for this purpose the more powerful mathematical calculus of Riemann, Ricci, and Levi-Civita.

In 1911, Einstein put forward the Principle of Equivalence, thus turning the subject towards gravitation for the first time. By postulating that not only mechanical but optical and electrical phenomena in a field of gravitation and in a field produced by acceleration of the observer were equivalent, he deduced the displacement of the spectral lines on the sun and the displacement of a star during a total eclipse. In the latter case, however, he predicted only the half-deflection, since he was still working with Newton's law of gravitation. Freundlich at once examined plates obtained at previous eclipses, but failed to find sufficient data; he also prepared to observe the eclipse of 1914 in Russia with this object, but was stopped by the outbreak of war. Another attempt was made by the Lick Observatory at the not very favourable eclipse of 1918. Only preliminary results have been published; according to the information given, the probable accidental error of the mean result (reduced to the sun's limb) was about $1''.6$, so that no conclusion was permissible.

The principle of equivalence opened up the possibility of a general theory of relativity not confined to uniform motion, for it pointed a way out of the objections which had been urged against such an extension from the time of Newton. At first the opening seemed a very narrow one, merely indicating that the objections could not be considered final until the possibilities of complications by gravitation had been more fully exhausted. By 1913, Einstein had surmounted the main difficulties. His theory in a complete form was published in 1915; but it was not generally accessible in England until a year or two later. As this theory forms the main subject-matter of the book, we may leave our historical survey at this point.

The Relativity of Time

A. S. Eddington, The Relativity of Time, *Nature* **106**, 802-804 (17 February 1921)

The philosopher discusses the significance of time; the astronomer measures time. The astronomer goes confidently about his business and does not think of asking the philosopher what exactly is this thing he is supposed to be measuring; nor does the philosopher always stop to consider whether time in his speculations is identical with the time which the world humbly accepts from the astronomer. In these circumstances it is not surprising that some confusion should have arisen.

In many globular clusters there are stars which oscillate in intrinsic brightness; let us select two such stars from different clusters and invite all the astronomers in the universe to measure the true interval of time between the moments of maximum light of the two stars. They must, of course, make whatever measurements and calculations they consider necessary to allow for the finite velocity of light. It may easily happen that the astronomers on Arcturus report that the two maxima were simultaneous; whereas those on the earth report an interval of *ten years* between the same two maxima. There is here no question of observational error; the recognised terrestrial method necessarily gives a discordant result when on Arcturus, owing to its different motion.

Our first impulse is to blame the astronomers. Evidently they are not giving us the true time-interval; and now that they are informed of the discordance they ought to give up their out-of-date procedure. But the astronomers reply: "Tell us, then, how we ought to find this 'true time'. By what characteristics are we to recognize it?" No answer has been given. Michelson and others sought in vain for an answer; for if our velocity through the aether could be defined, it would single out one universal system of time-measurement which might reasonably (if somewhat arbitrary) be called true. Meanwhile the phrase *true time* is a "meaningless noise." It is idle to contest with those who hold that the thing exists and ought to be regarded. "Who would give a bird the lie, though he cry 'Cuckoo' never so?"

The direction of Northampton measured by astronomers at Cambridge is due west; measured by astronomers at Greenwich it is north-west. It is no use to tell them that they must adopt a different plan, and find a "true direction" of Northampton which does not show these discordances. They reply: "We are perfectly aware that there must be discordances, as you call them; but that is in the nature of a relative property like direction; as for this true direction which shall be the same from all stations, we have no idea what you are talking about."

The time determined by astronomers and in general use is thus a fictitious time, or, in the usual phrase, it is *relative* to terrestrial observers. Similarly it has been found that extension in space is also relative. When the Copernican theory led to the abandonment of the geocentric view of the universe, the revolution did not go far enough; it was thought that we could pass to the heliocentric outlook by merely allowing for what in pure geometry would be called a change of origin. Actually a more profound transformation is necessary. For example, the Michelson-Morley experiment is a terrestrial experiment, but its theory is treated from a heliocentric point of view; that is to say, account is taken of the varying orbital motion of the earth; it finishes a proof of the famous FitzGerald contraction, and much ingenuity has been spent on an electrical explanation of this curious property of matter. Einstein's theory waves this aside with the remark: "Of course, your results appear strange when you describe the apparatus in terms of a space and time which do not belong to it. Your electromagnetic discussion is no doubt valid, but it is leading you away from the root of the matter; the immediate explanation lies in the difference between the heliocentric and geocentric space and time systems."

It was shown by Minkowski that all these fictitious spaces and times can be united in a single continuum of four dimensions. The question is often raised whether this four-dimensional space-time is real, or merely a mathematical construction; perhaps it is sufficient to reply that it can at any rate not be less real than the fictitious space and time which it supplants. Terrestrial observers divide the four-dimensional world into a series of sections or thin sheets (representing space) piled in an order which signifies time; in other words, the enduring universe is analyzed into a succession of instantaneous states. But this division is purely geometrical. The physical structure of the enduring world is not laminated in this way; and there is nothing to prevent another observer drawing his geometrical sections in a different direction. In fact, he will do so if his motion differs from ours. Now it may seem that we have been paying too much deference to the astronomers: "After all, they did not discover time. Time is something of which we are immediately conscious." I venture to differ and to suggest that (subject to certain reservations) time as now understood *was* discovered by an astronomer – Römer. By our sense of vision it appears to us that we are present at events far distant from us, so that they seem to occur in instants of which we are immediately conscious. Römer's discovery of the finite velocity of light has forced us to abandon that view; we still like to think of *world-wide* instants, but the location of distant events among them is a matter of hypothetical calculations, not of perception. Since Römer, time has become a mathematical construction devised to give the least disturbance to the old illusion that the instants in our consciousness are world-time.

Without using any external sense, we are conscious of the flight of time. This, however, is not a succession of world-wide states, but a succession of events at one place – not a pile of sheets, but a chain of points. Common-sense demands that this time-succession should be essentially different from the space-succession of points along a line. The preservation of a fundamental distinction between timelike succession and spacelike succession is essential in any acceptable theory. Thus in the four-dimensional world we recognize that there are two types of ordered succession of events which have no common measure; type A is like the succession of instants in our minds, and type B is the relation of order along a line in space. Proceeding from the instant "here-now", I can divide the regions of the world into two zones, according as they are reached by a succession of type A (my absolute past and future), or of type B (my absolute "elsewhere"). This scheme of structure is very different from the supposed laminated structure of the older view. Since we believe that this distinction of types A and B corresponds to something in the actual structure of the world, it is likely to determine the various natural phenomena that are observed. Thus it determines the propagation of light, since it is found that the line of a light-pulse is always on the boundary between the two zones above-mentioned. More important still, a particle of matter is a structure which can occupy a chain of points only of type A. Since we are limited by our material bodies, it must be this type of succession which we immediately experience; we are aware of the existence of the other type only by deduction from the indications of our external senses.

Objection is sometimes raised to the extravagantly important part taken by light-signals and light-propagation in Einstein's discussion of space and time. But Einstein did not invent a space and time depending on light-signals; he pointed out that the space and time already in general use depended on light-signals and equivalent processes, and proceeded to show the consequences of this. Turning from fictitious space and time to the absolute four-dimensional world, we still find the velocity of light playing a very prominent part. It is scarcely necessary to offer any excuse for this. Whether the substratum of phenomena is called *aether* or *world* or *space-time*, one requirement of its structure is that it should propagate light with this velocity.

The resolution of the four-dimensional continuum into a succession of instantaneous spaces is not dictated by anything in the structure of the continuum. Nevertheless, it is convenient, and corresponds approximately to our practical out-look on the world; and it is rarely necessary to go back to the undivided world. We have to go back to the undivided world when a comparison is made between the phenomena experienced by observers with different motions, who make the resolution in different directions. Moreover, a world-wide resolution into a space and time with the familiar properties is possible only when the continuum satisfies certain conditions. Are these conditions rigorously satisfied? They are not; that is Einstein's second great discovery. It is no more possible to divide the universe in this way than to divide the whole sky into squares. We have tried to make the division, and it has failed; and to cover up the consequences of the failure we have introduced an almost supernatural agency – gravitation. When we cease to strive after this impossibility - a mode of division which there was never any adequate reason for believing to be possible – gravitation as a separate agency becomes unnecessary. Our concern here is with the bearing of this result on time. The relative time for an observer is a construction extended by astronomers throughout the universe according to mathematical rules; but these rules break down in a region disturbed by the proximity of heavy matter, and cannot be fulfilled accurately. We can preserve our time-partitions only by making up fresh rules as we require them. The local time for a particular observer is always definite, and is the physical representation of the flight of instants of which he is immediately aware; the extended mesh-work of co-ordinates radiating from this is drawn so as to conform roughly to certain rules – so as not to violate too grossly certain requirements which the untutored mind thought necessary at one time. Subject to this, time is merely one of four co-ordinates, and its exact definition is arbitrary.

To sum up, world-wide time is a mathematical system of location of events according to rules which on examination can only be regarded as arbitrary; it has not any structural – and still less any metaphysical – significance. Local time, which for animate beings corresponds to the immediate time-sense, is a type of linear succession of events distinct from a pure spacelike succession; and this distinction is fully recognized in the relativity theory of the world.

"Space" or "Aether"?

A. S. Eddington, "Space" or "Aether"? *Nature* **107**, 201 (April 14, 1921)

Your readers are indebted to Mr. Bonacina's letter in NATURE of April 7 for a very clear statement of a fundamental point in the relativity controversy, and it is important that the views held with regard to it should be clearly understood. The issue is stated concisely in the sentence "the relativists seem now ... to indicate that space, instead of being conditioned by matter, is itself the foundation of matter and physical forces." Now it seems clear that if any relativist expresses himself in terms like these he cannot be regarding space as mere emptiness or as the arbitrary coordinate system of the pure mathematician; for him it is the substitution of matter, light and electric force – that is to say, it is the thing which most of us call aether. Since it is not matter, it has not (and we ought not to expect it to have) the material properties of density, elasticity or even velocity; but it has other dynamical attributes, measured by tensor expressions, which stand in much the same relation towards it that mass and strain do towards matter. It is, in short, a physical medium. It is sometimes stated that the relativity theory does away with the aether; the defence of this statement must be left to those who make it; I do not think it is the view of Prof. Einstein. It seems more reasonable to say that relativity has added to the importance of the aether by enlarging its functions.

But it must not be thought that the whole issue reduces to a question of terminology. It will naturally be asked: How can those who believe in a physical aether regard gravitation and electromagnetic phenomena as the "outcome of the geometry of the universe"? The phrase is Prof. Weyl's and reference to his book, "Raum, Zeit, Materie," shows that he believes in a physical aether, and does not mind saying so. "We shall use," he says the term "condition of the world-aether" as synonymous with "metric" in order to intimate the *real* character of the metric." We must recall that the geometric quantity called "distance" is none other than the material or aetherial attribute of "extencion" as Mr. Bonacina admits. Thus experimental geometry, which comprises the study of distances, is the science of the aether so far as its attribute of extension is concerned. The sentence then means that not only the phenomena immediately recognized as spatial, but also mechanical and electrical phenomena, fall into place in a complete development of the theory of extension – a truly remarkable discovery. They do not introduce any other attribute of the aether. I think it is because physical science is confined to this one attribute of the substratum of the universe that such qualities as beauty lie outside its scope.

The statement that the phenomena of mechanics are the outcome of the geometry of the world implies the complementary statement that the phenomena of experimental geometry are the outcome of the mechanics of the world. Either form expresses the central truth of the generalized relativity theory, but the great advance lies not so much in the conception of the idea as in the discovery of the key to this unification of geometry and mechanics. The unification leaves us with a redundancy of names, and apparently there is some divergence of view as to the right name for the fundamental substratum of everything. Since it is the medium the condition of which determined light and electromagnetic force, we may call it *aether*; since it is the subject-matter of the science of geometry, we may call it *space*; sometimes in order to avoid giving preference to either aspect, it is called by Minkowski's term *world.*

A.S. Eddington

Observatory, Cambridge, April 11

INDEX

Made in the USA
Lexington, KY
07 April 2018